Understanding Digital Image Processing

Vipin Tyagi

Department of Computer Science and Engineering
Jaypee University of Engineering and Technology
Raghogarh, Guna (MP), India

CRC Press
Taylor & Francis Group
Boca Raton London New York

CRC Press is an imprint of the
Taylor & Francis Group, an **informa** business
A SCIENCE PUBLISHERS BOOK

Cover photo credits: Prahlad Bubbar collection

MATLAB® and Simulink® are trademarks of The MathWorks, Inc. and are used with permission. The MathWorks does not warrant the accuracy of the text or exercises in this book. This book's use or discussion of MATLAB® and Simulink® software or related products does not constitute endorsement or sponsorship by The MathWorks of a particular pedagogical approach or particular use of the MATLAB® and Simulink® software.

CRC Press
Taylor & Francis Group
6000 Broken Sound Parkway NW, Suite 300
Boca Raton, FL 33487-2742

First issued in paperback 2021

© 2018 by Taylor & Francis Group, LLC
CRC Press is an imprint of Taylor & Francis Group, an Informa business

No claim to original U.S. Government works

Version Date: 20180730

ISBN-13: 978-0-367-78082-1 (pbk)
ISBN-13: 978-1-138-56684-2 (hbk)

**Visit the Taylor & Francis Web site at
http://www.taylorandfrancis.com**

**and the CRC Press Web site at
http://www.crcpress.com**

Preface

◇◇◇

Digital Images are an integral part of our digital life. Digital images are used in every aspect of or daily life. Digital image processing concerns with techniques to perform processing on an image, to get an enhanced image or to extract some useful information from it to make some decisions based on it. Digital image processing techniques are growing at a very fast speed. This book, *Understanding Digital Image Processing* aims on providing digital image processing concepts in a simple language with the help of examples. The major objective of this text book is to introduce the subject to the readers and motivate for further research in the area.

To explain the concepts, MATLAB® functions are used throughout the book. MATLAB® Version 9.3 (R2017b), Image Acquisition Toolbox Version 5.3 (R2017b), Image Processing Toolbox, Version 10.1 (R2017b) are used to create the book material. My thanks to MathWorks for providing support in preparation of this book.

The functions provided by Image Processing Toolbox™ are given in Appendix A. The details of these functions are available at on the website https://in.mathworks.com/help/images/index.html

For MATLAB® product information, please contact:

The MathWorks, Inc.

3 Apple Hill Drive

Natick, MA, 01760-2098 USA

Tel: 508-647-7000

Fax: 508-647-7001

E-mail: info@mathworks.com

Web: mathworks.com

How to buy: www.mathworks.com/store

'C' language is a very popular programming language. Various functions written in 'C' language are given in Appendix B. My sincere thanks to Mr. Dwayne Phillips for permitting to use the code written by him. Related sub-functions and their descriptions are available at http://homepages.inf.ed.ac.uk/rbf/BOOKS/PHILLIPS/.

A glossary of common image processing terms is provided in Appendix C.

A bibliography of the work in the area of image processing is also given at end of the book. We are thankful to all authors whose work has been used in preparation of the manu-script. We have tried to include all such work in bibliography, but if we have skipped some work by error, we will include in next version of the book, Color version of some figures have been provided at the end of the book. These figures also have corresponding black & white versions in the text.

The target audience spans the range from the undergraduate with less exposure to subject to research students interested in learning digital image processing. Many texts are available in the area of digital image processing. In this book, objective is to explain the concepts in a very simple and understandable manner. Hope this book will succeed in its aim.

This work would not have been possible without the help and mentoring from many,in particular, my teacher Prof. Vinod K. Agarwal, Meerut. Special thanks to my dear scholars Mr. K. B. Meena, Dr. Ashwani Kumat and Dr. Deepshikha Tiwari, for their help and support in preparation of the manuscript.

The research work of several researchers contributed to a substantial part of some sections of the book. I thankfully acknowledge their contributions.

It has been a pleasure working with Taylor and Francis Publishers in the development of the book. Thanks to Mr. Vijay Primlani for his kind and timely support in publishing the book and for handling the publication.

Contents

1

Introduction to Digital Image Processing

1.1 Introduction

In today's digital life, digital images are everywhere around us. An image is a visual representation of an object, a person, or a scene. A digital image is a two-dimensional function f(x, y) that is a projection of a 3-dimesional scene into a 2-dimensional projection plane, where x, y represents the location of the picture element or pixel and contains the intensity value. When values of x, y and intensity are discrete, then the image is said to be a digital image. Mathematically, a digital image is a matrix representation of a two-dimensional image using a finite number of points cell elements, usually referred to as pixels (picture elements, or pels). Each pixel is represented by numerical values: for grayscale images, a single value representing the intensity of the pixel (usually in a [0, 255] range) is enough; for color images, three values (representing the amount of red (R), green (G), and blue (B)) are stored. If an image has only two intensities, then the image is known as a binary image. Figure 1.1 shows a color image and its red, green and blue components. The color image is a combination of these three images. Figure 1.1e shows the 8-bit grayscale image corresponding to color image shown in Fig. 1.1a. Figure 1.1 also shows the matrix representation of a small part of these images.

MATLAB® supports the following image types:

1. **Grayscale:** A grayscale image, having M×N pixels is represented as a matrix of double datatype of M×N size in MATLAB. Element values denote the pixel grayscale intensities in the range [0,1] with 1 = white and 0 = black.
2. **True-color RGB:** A true-color red-green-blue (RGB) image is represented as a three-dimensional M×N×3 double matrix in MATLAB. Each pixel has

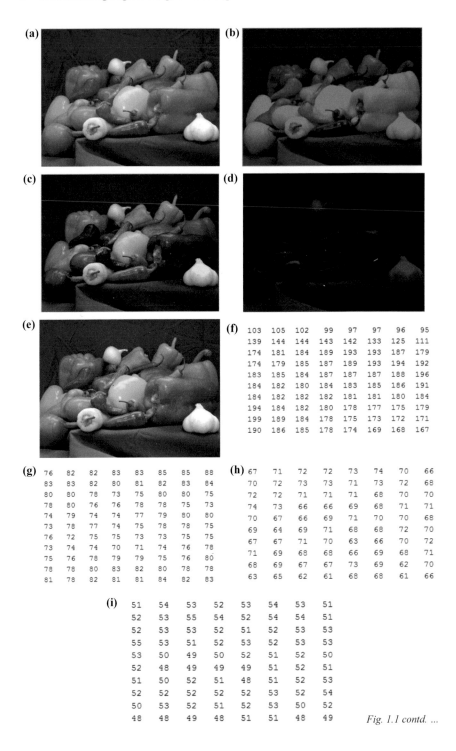

Fig. 1.1 contd. ...

red, green and blue components. The color components of a pixel (m,n) are denoted as (m,n,1) = red, (m,n,2) = green, (m,n,3) = blue.

3. **Indexed:** In MATLAB, Indexed (paletted) images are represented with an index matrix of size M×N and a colormap matrix of size K×3. The colormap matrix holds all colors used in the image and the index matrix represents the pixels by referring to colors in the colormap.

4. **Binary:** In MATLAB, a binary image having two values, 1 (white) or 0 (black), is represented by an M×N logical matrix.

5. **uint8:** In MATLAB, this type uses less memory and some operations compute faster than with double types.

In image processing operations, most of the operations are performed in grayscale images. For color image processing applications, a color image can be decomposed into Red, Green and Blue components and each component is processed independently as a grayscale image. For processing, an indexed image is converted to grayscale image or RGB color image for most operations.

MATLAB commands imread and imwrite are used for reading and writing image files as:

I = imread(filename);

imwrite(I, filename) writes image data I to the file specified by filename, inferring the file format from the extension. The bit depth of the output image depends on the data type of I and the file format. For most formats:

- If I is of data type uint8, then imwrite outputs 8-bit values.
- If I is of data type uint16 and the output file format supports 16-bit data (JPEG, PNG and TIFF), then imwrite outputs 16-bit values. If the output file format does not support 16-bit data, then imwrite returns an error.
- If I is a grayscale or RGB color image of double or single data type, then imwrite assumes that the dynamic range is [0,1] and automatically scales the data by 255 before writing it to the file as 8-bit values. If the data in I is single, convert I to double before writing to a GIF or TIFF file.
- If I is of logical data type, then imwrite assumes that the data is a binary image and writes it to the file with a bit depth of 1, if the format allows it. BMP, PNG, or TIFF formats accept binary images as input arrays.

...Fig. 1.1 contd.

Fig. 1.1. (a) A color image; (b) Red Component of color image (a); (c) Green Component of color image (a); (d) Blue Component of color image (a); e) Color image (a) converted into 8-bit grayscale image; (f) Matrix representation of upper left corner of image (b); (g) Matrix representation of upper left corner of image (c); (h) Matrix representation of upper left corner of image (d); (i) Matrix representation of upper left corner of image (e).

Original image reprinted with permission of The MathWorks, Inc.

(For color images of Fig. 1.1(a), (b), (c), (d) see Color Figures Section at the end of the book)

In literature, the following three levels of image processing operations are defined:

- Low-Level image processing: Primitive operations on images (e.g., contrast enhancement, noise reduction, etc.) are under this category, where both the input and the output are images.
- Mid-Level image processing: In this category, operations involving extraction of attributes (e.g., edges, contours, regions, etc.), from images are included.
- High-Level image processing: This category involves complex image processing operations related to analysis and interpretation of the contents of a scene for some decision making.

Image processing involves many disciplines, mainly computer science, mathematics, psychology and physics. Other areas, such as artificial intelligence, pattern recognition, machine learning, and human vision, are also involved in image processing.

1.2 Typical Image Processing Operations

Image processing involves a number of techniques and algorithms. The most representative image processing operations are:

- **Binarization:** Many image processing tasks can be performed by converting a color image or a grayscale image into binary in order to simplify and speed up processing. Conversion of a color or grayscale image to a binary image having only two levels of gray (black and white) is known as binarization.
- **Smoothing:** A technique that is used to blur or smoothen the details of objects in an image.
- **Sharpening:** Image processing techniques, by which the edges and fine details of objects in an image are enhanced for human viewing, are known as sharpening techniques.
- **Noise Removal and De-blurring:** Before processing, the amount of noise in images is reduced using noise removal filters. Image removal technique can sometimes be used, depending on the type of noise or blur in the image.
- **Edge Extraction:** To find various objects before analyzing image contents, edge extraction is performed.
- **Segmentation:** The process of dividing an image into various parts is known as segmentation. For object recognition and classification segmentation is a pre-processing step.

1.3 History of Digital Image Processing

Earlier digital image processing was mainly used in the newspaper industry to make images look better or for simply converting black and white images into color images. In the 1920s, digital images were transmitted electronically between London and New York. Initial Bartlane cable picture systems were capable of coding an image using five gray levels; this was later enhanced to 15 gray levels in 1929. Actual digital image processing started after the invention of digital computers and related technologies, including image storage, display and transmission. In the 1960s, powerful computers gave birth to meaningful digital image processing. Satellite Images of the moon, taken by Ranger 7 U.S. spacecraft, were processed at the Jet Propulsion laboratory at California. At the same time, use of digital image processing began in various activities relating to astronomy, medical image processing, remote sensing, etc. From 1960 onwards, the use of digital image processing techniques has grown tremendously. These techniques are now used in almost every part of our life. They have applications in the fields of defence, astronomy, medicine, law, etc.

Some common examples of digital image processing are fingerprint recognition, processing of satellite images, weather prediction, character recognition, face recognition, product inspection and assembly.

1.4 Human Visual System

The human visual system consists of two parts: eye and brain. The human eye acts as a receptor of images by capturing light and converting it into signals. These signals are then transmitted to the brain for further analysis. Eyes and brain work in combination to form a picture.

The human eye is analogous to a camera. The structure of human eye is shown in Fig. 1.2:

Various parts of the human eye are identified:

- Primary Lens: Cornea and aqueous humour, used to focus incoming light signal.
- Iris: The iris dynamically controls the amount of light entering the eye, so that the eye can function in various lighting conditions, from dim light to very bright light. The portion of the lens not covered by the iris is called the pupil.
- Zonula: This is a muscle that controls the shape and positioning (forward and backward) of the eye's lens.

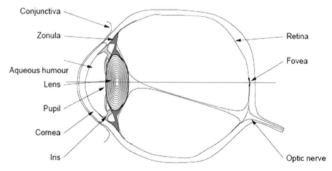

Fig. 1.2. Human eye.

- Retina: provides a photo-sensitive screen at the back of the eye; it converts the light hitting the retina into nerve signals.
- Fovea: A small central region of the retina that contains a large number of photo-sensitive cells and provides very good resolution.
- Optic nerve: These nerves carry the signals generated by the retina to the brain.

Human vision has a "blind spot" in the area of the retina where the optic nerves are attached. This blind spot does not have any photosensitive cells.

Light sensitive cells in the brain are of two types: rods and cones. There are about 120 million rod cells and 6 million cone cells. Rod cells provide visual capability at very low light levels and are very sensitive. Cone cells provide our daytime visual facilities and perform best at normal light levels.

Rods and cones are not uniformly distributed across the retina. The cones are concentrated in the center, while the rods are away from the center (Fig. 1.3). A yellow spot (macula), of size 2.5 to 3 mm in diameter, is found in the middle of the retina. Cones are very tightly packed together and the blood vessels within the fovea and other cells are pulled aside in order to expose them directly to the light.

In dim light, such as moonlight, rods in our eyes are activated and the fovea effectively acts as a second blindspot. To see small objects at night, one must shift the vision slightly to one side, around 4 to 12 degrees, so that the light falls on some rods.

Although there are about 120 million rods and 6 million cone cells in the retina, there are less than a million optic nerve fibres which connect them to the brain. This means that there cannot be a single one-to-one connection between the photoreceptors and the nerve fibres. The number of receptors connecting to each fibre is location dependent.

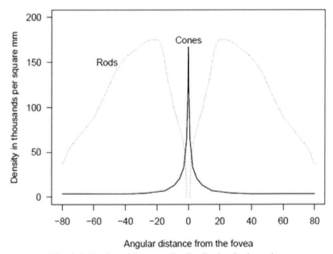

Fig. 1.3. Rods and Cones cells distribution in the retina.

In the fovea, cones have a one-to-one interaction, while in the outer part of the retina, about 600 rods are connected to each nerve fibre.

1.5 Classification of Digital Images

With regard to the manner in which they are stored, digital images can be classified into two categories: (1) Raster or Bitmap image, (2) Vector image.

A bitmap or raster image is a rectangular array of sampled values or pixels. These images have a fixed number of pixels. In the zooming of a raster image, mathematical interpolation is applied. The quality of a zoomed image degrades after a particular value of zooming factor, as shown in Fig. 1.4.

The resolution of a bitmap image is determined by the sensing device. BMP, GIF, PNG, TIFF and JPEG common image formats are bitmap or raster image formats.

On the other hand, vector images are stored in the form of mathematical lines and curves. Information like length, color, thickness, etc., is stored in the form of a vector. These images can be displayed in any size, any resolution and on any output

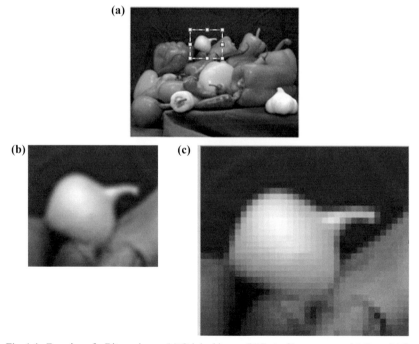

Fig. 1.4. Zooming of a Bitmap image (a) Original image (b) Part of image zoomed 4 times (c) Same part of the image zoomed 8 times.
Original image reprinted with permission of The MathWorks, Inc.
(For color image of Fig. 1.4(a), see Color Figures Section at the end of the book)

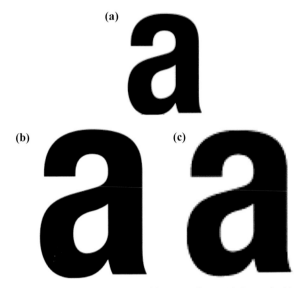

Fig. 1.5. (a) An image (b) Image in (a) zoomed in vector format (c) Image in (a) zoomed in Raster format.

device. Vector images are suitable in illustrations, line art, font, etc. The difference in zooming of a vector image and raster image can be observed in Fig. 1.5. The degradation in the quality due to zooming is clearly visible on the boundaries of the character stored in raster format.

1.6 Digital Image File Types

There are a number of digital image file types available these days. The most commonly used image file types are: JPEG, GIF, TIFF, PNG and BMP. Image file types are based on the compression technique used for reducing the size of the image file. Images in various file types may differ in color, if color has been used. An image in its simplest form may contain only two intensities, i.e., black and white, and needs only 1 bit to represent intensity at each pixel.

- **TIFF** (Tagged Image File Format): This format is a very flexible and may be based on a lossy or lossless compression technique. The details of the compression technique are stored in the image itself. Generally, TIFF files use a lossless image storage format and, hence, are quite large in size.
- **Portable Network Graphics** (PNG): This format is a lossless storage format and uses patterns in the image to compress the image. The compression in PNG files is exactly reversible, meaning the uncompressed image is identical to the original image.
- **Graphical Interchange Format** (GIF): This format creates a table of upto 256 colors from 16 million colors. If the image to be compressed has less

than 256 colors, then the GIF image has exactly the same color. But if the number of colors is greater than 256, then the GIF approximates the colors in the image using a table of the 256 colors that are available.

- **Joint Picture Experts Group** (JPG or JPEG): This format is an optimized format for photographs and continuous tone images that contain a large number of colors. JPEG files can achieve high compression ratios while maintaining clear image quality.
- **RAW:** This is a lossless image format available on some digital cameras. These files are very small but the format is manufacturer dependent, therefore, the manufacturer's software is required in order to view the images.
- **Bitmapped Image** (BMP) is an uncompressed proprietary format invented by Microsoft.

Some other common image file formats are PSD, PSP, etc., which are proprietary formats used by graphics programs. Photoshop's files have the PSD extension, while Paintshop Pro files use the PSP extension.

MATLAB supports a number of image file formats, e.g., Windows Bitmap, Windows Cursor resources, Flexible Image Transport, Graphics Interchange Format, Windows Icon resources, JPEG, Portable Bitmap, Windows Paintbrush, Tagged Image File Format.

The details can be obtained using MATLAB function imformats. The imformats function can be used if the file format exists in the registry as:

formatStruct = imformats ('bmp'), then the output will look like:
format Struct = struct with fields:
  ext: {'bmp'}
  isa: @isbmp
  info: @imbmpinfo
  read: @readbmp
  write: @writebmp
  alpha: 0
  description: 'Windows Bitmap'

1.7 Components of an Image Processing System

A complete digital image processing system comprises of various elements, such as image acquisition, storage, image processing, displays, etc. (Fig. 1.6).

Sensing devices are used to capture the image. The sensing device senses the energy radiated by the object and converts it into digital form. For example, a digital camera senses the light intensity and converts into the digital image form. Image processing elements are used to perform various operations on a digital image. It requires a combination of hardware and software. Storage is a very important part of an image processing system. The size of an image or video file is very large. For instance, an 8-bit image having 1024 x 1024 pixels requires 1 megabyte of storage space. Therefore, mass storage devices are required in image processing systems.

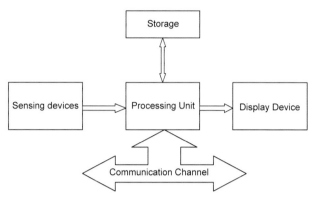

Fig. 1.6. A digital image processing system.

Display devices are required to display the images. These can be a computer monitor, mobile screen, projector for display or hardcopy devices, such as printers. A communication channel is also essential for sending and receiving images.

1.8 Applications of Digital Image Processing

Digital image processing techniques are now used in a number of applications; some common applications are given below.

In medicine: Several medical tools use image processing for various purposes, such as image enhancement, image compression, object recognition, etc. X-radiation (X-rays), computed tomography scan (CT scan), positron-emission tomography (PET), Single-photon emission computed tomography (SPECT), nuclear magnetic resonance (NMR) spectroscopy and Ultra-Sonography are some popular pieces of medical equipment based on image processing.

In agriculture: Image processing plays a vital role in the field of agriculture. Various paramount tasks such as weed detection, food grading, harvest control and fruit picking are done automatically with the help of image processing. Irrigated land mapping, determination of vegetation indices, canopy measurement, etc., are possible with good accuracy through the use of imaging techniques in various spectrums, such as hyper spectral imaging, infrared, etc.

In weather forecasting: Image processing also plays a crucial role in weather forecasting, such as prediction of rainfall, hailstorms, flooding. Meteorological radars are widely used to detect rainfall clouds and, based on this information, systems predict immediate rainfall intensity.

In photography and film: Retouched and spliced photos are extensively used in newspapers and magazines for the purpose of picture quality enhancement. In movies, many complex scenes are created with image and video editing tools which are based on image and video processing operations. Image processing-based methods are used to predict the success of upcoming movies. For a global media and entertainment company, Latent View extracted over 6000 movie posters from IMDB along with their metadata (genre, cast, production, ratings, etc.), in order to predict the movies' success using image analytics. The colour schemes and objects in the movie posters were analyzed using Machine Learning (ML) algorithms and image processing techniques.

In entertainment and social media: Face detection and recognition are widely used in social networking sites where, as soon as a user uploads the photograph, the system automatically identifies and gives suggestion to tag the person by name.

In security: Biometric verification systems provide a high level of authenticity and confidentiality. Biometric verification techniques are used for recognition of humans based on their behaviours or characteristics. To create alerts for particularly undesirable behaviour, video surveillance systems are being employed in order to analyze peoples' movements and activities. Several banks and other departments are using these image processing-based video surveillance systems in order to detect undesired activities.

In banking and finance: The use of image processing-based techniques is rapidly increasing in the field of financial services and banking. 'Remote deposit capture' is a banking facility that allows customers to deposit checks electronically using mobile devices or scanners. The data from the check image is extracted and used in place of a physical check. Face detection is also being used in the bank customer authentication process. Some banks use 'facial-biometric' to protect sensitive information. Signature verification and recognition also plays a significant role in authenticating the signature of the customers. However, a robust system used to verify handwritten signatures is still in need of development. This process has many challenges because handwritten signatures are imprecise in nature, as their corners are not always sharp, lines are not perfectly straight, and curves are not necessarily smooth.

In marketing and advertisement: Some companies are using image-sharing through social media in order to track the influence of the company's latest products/advertisement. The tourist department uses images to advertise tourist destinations.

In defence: Image processing, along with artificial intelligence, is contributing to defence based on two fundamental needs of the military: one is autonomous operation and the other is the use of outputs from a diverse range of sophisticated sensors for predicting danger/threats. In the Iran-Iraq war, remote sensing technologies were employed for the reconnaissance of enemy territory. Satellite images are analyzed in order to detect, locate and destroy weapons and defence systems used by enemy forces.

In industrial automation: An unprecedented use of image processing has been seen in industrial automation. The 'Automation of assembly lines' system detects the position and orientation of the components. Bolting robots are used to detect the moving bolts. Automation of inspection of surface imperfection is possible due to image processing. The main objectives are to determine object quality and detect any abnormality in the products. Many industries also use classification of products by shape automation.

In forensics: Tampered documents are widely used in criminal and civil cases, such as contested wills, financial paper work and professional business documentation. Documents like passports and driving licenses are frequently tampered with in order to be used illegally as identification proof. Forensic departments have to identify the authenticity of such suspicious documents. Identifying document forgery becomes increasingly challenging due to the availability of advanced document-editing tools. The forger uses the latest technology to perfect his art. Computer scan documents are copied from one document to another to make them genuine. Forgery is not only confined to documents, it is also gaining popularity in images. Imagery has a remarkable role in various areas, such as forensic investigation, criminal investigation, surveillance systems, intelligence systems, sports, legal services, medical imaging, insurance claims and journalism. Almost a decade ago, Iran was accused of doctoring an image from its missile tests; the image was released on the official website, Iran's Revolutionary Guard, which claimed that four missiles were heading skyward simultaneously. Almost all the newspaper and news magazine published this photo including, The Los Angeles Times, The Chicago Tribune and BBC News. Later on, it was revealed that only three missiles were launched successfully, one missile failed. The image was doctored in order to exaggerate Iran's military capabilities.

In underwater image restoration and enhancement: Underwater images are often not clear. These images have various problems, such as noise, low contrast, blurring, non-uniform lighting, etc. In order to restore visual clarity, image enhancement techniques are utilized.

Summary

- Digital images are very important in today's digital life.
- An image may be a grayscale image or a color image. A grayscale image, having only two intensities, is termed as a binary image.
- Digital images can be classified as Raster image or Vector image, with regard to the manner in which the image is stored.
- Some common image file types which are used in our daily life are JPEG, GIF, TIFF, PNG and BMP.
- The main components of an image processing system are (1) Sensing device, (2) Image processing elements, (3) Storage device and (4) Display device.

2

Digital Image Representation

2.1 Digital Image

A digital image is made up of $M \times N$ pixels, and each of these pixels is represented by k bits. A pixel represented by k bits can have 2^k different shades of gray in a grayscale image. These pixel values are generally integers, having values varying from 0 (black pixel) to 2^k-1 (white pixel). The number of pixels in an image defines resolution and determines image quality.

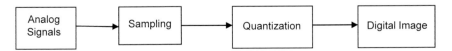

Fig. 2.1. Digital image formation process.

2.2 Sampling and Quantization

When an image is sensed and converted into digital form, the image is discretized in two ways: sampling and quantization. The concepts of sampling and quantization can be understood by following example in Fig. 2.2, in which a one-dimensional signal is quantized on a uniform grid. The signal is sampled at ten positions ($x = 0,..., 9$), and each sampled value is then quantized to one of seven levels ($y = 0,..., 6$). It can be observed that the samples collectively describe the gross shape of the original signal but that smaller variations and structures may not be represented, i.e., information may have been lost mathematically. The digital image is an approximation of a real image and the quality depends on the number of samples taken and the quantization.

Sampling is the process of converting coordinate values in digital form and Quantization is the process of digitizing amplitude values. The quality of an image and display device is measured in terms of resolution that depends on the sampling

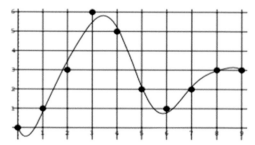

Fig. 2.2. A 1-dimensional signal quantized on a uniform grid.

and quantization process. The number of samples taken in digitization is known as image resolution and the quantization decides the accuracy of pixel intensities. The number of intensity values or pixels in an image is known as spatial resolution (Fig. 2.3) and the number of various graylevel values in an image is known as graylevel resolution (Fig. 2.4). These two factors also determine the space required to store an image, as can be observed from Table 2.1. If in an image, there is not a sufficient number of graylevels, then the smooth parts in the image exhibit 'false contouring'. This effect is clearly visible in Fig. 2.4.

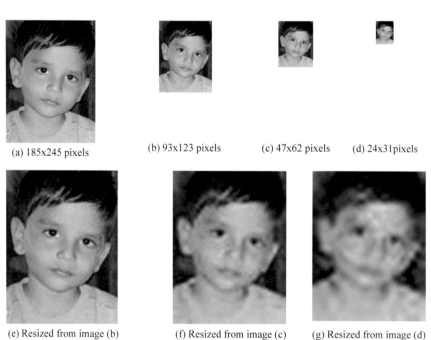

(a) 185x245 pixels   (b) 93x123 pixels   (c) 47x62 pixels   (d) 24x31pixels

(e) Resized from image (b)   (f) Resized from image (c)   (g) Resized from image (d)

Fig. 2.3. The effect of spatial resolution on the image quality.

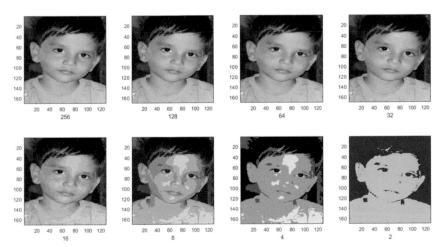

Fig. 2.4. The effect of gray level resolution on the image quality.

Table 2.1. Storage Requirement (in bytes) for an image with different parameter values.

Resolution	Bits per pixel				
	1	2	4	6	8
32 × 32	128	256	512	768	1,024
64 × 64	512	1,024	2,048	3,072	4,096
128 × 128	2,048	4,096	8,192	12,288	16,384
256 × 256	8,192	16,384	32,768	49,152	65,536
512 × 512	32,768	65,536	131,072	196,608	262,144
1024 × 1024	131,072	262,144	524,288	786,432	1,048,576

2.3 Color Models

Visible light is composed of various frequencies, approximately between 400 and 700 nm in the electromagnetic energy spectrum. Newton discovered that if a white sunlight beam is passed through a glass prism, the beam is divided into a continuous spectrum of colors ranging from violet to red (Fig. 2.5). The colors in this color spectrum are VIBGYOR (Violet, Indigo, Blue, Green, Yellow, Orange, Red). In this spectrum, the red color has the longest wavelength, and the violet color has the shortest wavelength. Wavelengths of various colors are given in Table 2.2.

The colors are observed by the human eye according to the wavelength of the light reflected by the object. For example, a yellow object reflects light having wavelength 577–597 nm, and absorbs most of the energy at other wavelengths. Similarly, an object appears white if it reflects relatively balanced light that is in all visible wavelengths.

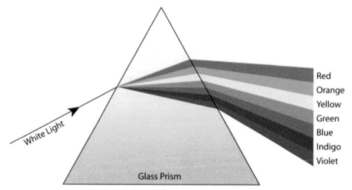

Fig. 2.5. Color spectrum after passing through a Prism.
(For color image of this Figure see Color Figures Section at the end of the book)

Table 2.2. Wavelength of various colors.

Color	Wavelength (nm)
Violet	380–435
Indigo	~ 446
Blue	435–500
Green	500–565
Yellow	565–590
Orange	590–625
Red	625–740

According to color theory, color formation is based on two processes: (i) Additive color formation and (ii) Subtractive color formation.

2.3.1 RGB Color Model

Additive colors are formed by combining spectral light in varying combinations. For example, computer screen produces color pixels by generating red, green, and blue electron guns at phosphors on the monitor screen. Additive color formation begins with black and ends with white and, as more color is added, the result becomes lighter and tends towards white (Fig. 2.6). Primary additive colors are Red, Green and Blue (R, G, B). The CIE (Commission Internationale de l'Eclairage - International Commission on Illumination) has standardized the wavelength values to the primary colors, which are as follows:

Red $(R) = 700.0$ nm
Green $(G) = 546.1$ nm
Blue $(B) = 435.8$ nm

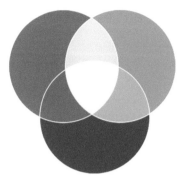

Fig. 2.6. Additive color system.
(For color image of this Figure see Color Figures Section at the end of the book)

In the RGB color model, the primary colors are represented as:

Red = (1,0,0), Green = (0,1,0), Blue = (0,0,1).
and the secondary colors of RGB are represented as
Cyan = (0,1,1), Magenta = (1,0,1), Yellow = (1,1,0).

The RGB model can be represented in the form of a color cube (Fig. 2.7). In this cube, black is at the origin (0,0,0), and white is at the corner, where $R = G = B = 1$. The grayscale values from black to white are represented along the line joining these two points. Thus a grayscale value can be represented as (x,x,x) starting from black = (0,0,0) to white = (1,1,1).

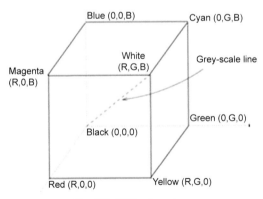

Fig. 2.7. RGB color cube.

2.3.2 CMY Color Model

In subtractive color formation, the color is generated by light reflected from its surroundings and the color does not emit any light of its own. In the subtractive colour system, black color is produced by a combination of all the colors (Fig. 2.8). For example, in printing, black color is generated by mixing all colors,

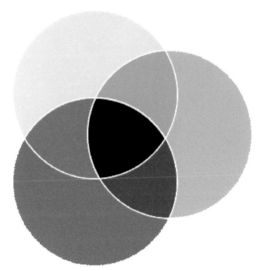

Fig. 2.8. Subtractive color system.
(For color image of this Figure see Color Figures Section at the end of the book)

and, hence, subtractive color method is used. In subtractive color mixing, white means no color and black is a combination of all colors. As more colors are added, the result becomes darker and tends towards black. Cyan, magenta and yellow (CMY) colors correspond to primary colors in subtractive color system.

The primary colors are added to generate the secondary colors yellow (red + green), cyan (green + blue), magenta (red + blue). Several other color models, such as YUV, YIQ and HSI also exist and can be used as per the requirement.

The RGB to CMY conversion can be performed using:

C=1-R

M=1-G   (2.1)

Y=1-B

In the CMY model, a pixel of color cyan, for example, reflects all other RGB colors but red. A pixel with the color of magenta, on the other hand, reflects all other RGB colors but green. Now, if cyan and magenta colors are mixed, blue will be generated, rather than white as in the additive color system. In the CMY model, the black color generated by combining cyan, magenta and yellow is not very impressive, therefore a new color, black, is added in CMY color model, generating CMYK model. In publishing industry this CMYK model is referred as four-color printing.

There is another popular color model known as the YUV model. In the YUV color model, Y is luminance and U and V represent chrominance or color information. The YUV color model is based on color information along with brightness information. The components of YUV are:

$$Y = 0.3\ R + 0.6\ G + 0.1\ B$$
$$U = B\text{-}Y \qquad\qquad (2.2)$$
$$V = R\text{-}Y$$

The luminance component can be considered as a grayscale version of the RGB image. There are certain advantages of YUV compared to RGB, which are:

- The brightness information is separated from the color information.
- The correlations between the color components are reduced.
- Most of the information is in the Y component, while the information content in the U and V is less.

The latter two properties are very useful in various applications, such as image compression. This is because the correlations between the different color components are reduced, thus, each of the components can be compressed separately. Besides that, more bits can be allocated to the Y component than to U and V. The YUV color system is adopted in the JPEG image compression standard.

2.3.3 YIQ Color Model

The YIQ color model takes advantage of the human visual system which is more sensitive to luminance variations in comparison to variance in hue or saturation. In the YIQ color system, Y is the luminance component, and I and Q represent chrominance U and V similar to YUV color systems. The RGB to YIQ conversion can be done using

$$Y = 0.299\ R + 0.587\ G + 0.114\ B$$
$$I = 0.596\ R\text{--}0.275\ G\text{--}0.321\ B \qquad\qquad (2.3)$$
$$Q = 0.212\ R\text{--}0.523\ G + 0.311\ B$$

The YIQ color model can also be described in terms of YUV:

$$Y = 0.3\ R + 0.6\ G + 0.1\ B$$
$$I = 0.74\ V\text{--}0.27\ U \qquad\qquad (2.4)$$
$$Q = 0.48\ V + 0.41\ U$$

2.3.4 HSI Color Model

The HSI model is based on hue (H), saturation (S), and intensity (I). Hue is an attribute associated with the dominant wavelength in a mixture of light waves, i.e., the dominant color as perceived by a human. Saturation refers to the relative purity of the amount of white light mixed with the hue. Intensity corresponds to the luminance component (*Y*) of the YUV and YIQ models. The advantages of HSI are:

- The intensity is separated from the color information (similar to YUV and YIQ models).
- The hue and saturation components are intimately related to the way in which human beings perceive color.

whereas RGB can be described by a three-dimensional cube, the HSI model is represented as a color triangle, as shown in Fig. 2.9.

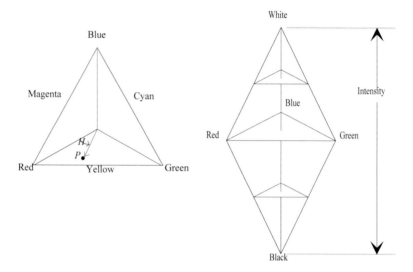

Fig. 2.9. HSI color triangle.

All colors lie inside the triangle whose vertices are defined by the three basic colors, red, green and blue. If a vector is drawn from the central point of the triangle to the color point *P*, then hue (*H*) is the angle of the vector with respect to the red axis. For example, 0° indicates red color, 60° yellow, 120° green, and so on. Saturation (*S*) is the degree to which the color is undiluted by white and is proportional to the distance to the center of the triangle.

The RGB to HSI conversion can be performed as follows:

$$H = \frac{1}{360°} \cdot \cos^{-1}\left[\frac{\frac{1}{2} \cdot \left[(R-G)+(R-B)\right]}{\sqrt{(R-G)^2 + (R-B)(G-B)}}\right], \text{ if } B \leq G$$

$$H = 1 - \frac{1}{360°} \cdot \cos^{-1}\left[\frac{\frac{1}{2} \cdot \left[(R-G)+(R-B)\right]}{\sqrt{(R-G)^2 + (R-B)(G-B)}}\right], \text{ otherwise}$$

$$S = 1 - \frac{3}{(R+G+B)} \cdot \min(R,G,B)$$

$$I = \frac{1}{3} \cdot (R+G+B)$$

(2.5)

2.4 Basic Relationships between Pixels

The neighboring relationship between pixels is very important in an image processing task. A pixel p(x,y) has four neighbors (two vertical and two horizontal neighbors) specified by [(x+1, y), (x–1, y), (x, y+1), (x, y–1)].

This set of pixels is known as the 4-neighbors of P, and is denoted by $N_4(P)$. All these neighbors are at a unit distance from P.

A pixel p(x,y) has four diagonal neighbors specified by, [(x+1, y+1), (x+1, y–1), (x–1, y+1), (x–1, y–1)].

This set of pixels is denoted by $N_D(P)$. All these neighbors are at a distance of 1.414 in Euclidean space from P.

The points $N_D(P)$ and $N_4(P)$ are together known as 8-neighbors of the point P, denoted by $N_8(P)$. In 8-neighbors, all the neighbors of a pixel p(x,y) are:

[(x+1, y), (x–1, y), (x, y+1), (x, y–1), (x+1, y+1), (x+1, y–1), (x–1, y+1), (x–1, y–1)].

Figure 2.10 shows 4-neigbours, diagonal neighbors and 8-neigbours of a pixel. For a pixel on the boundary, some neighboring pixels of a pixel may not exist in the image.

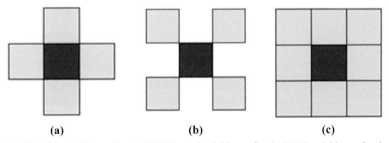

(a)   (b)   (c)

Fig. 2.10. (a) 4-neighbors of a pixel (b) Diagonal neighbors of a pixel (c) 8-neighbors of a pixel.

2.5 Adjacency

Adjacency of two pixels is defined in the terms of neighbors of a pixel. Two pixels p and q are said to be 4-adjacent if these pixels are 4-neighbors to each other, i.e., if these pixels have the same value in a binary image and q is 4-neighbor of p. Similarly, two pixels p and q are 8-adjacent, if these pixels are 8-neighbors to each other, i.e., if these pixels have the same value in a binary image and q is 8-neighbor of p. This concept can also be extended to grayscale images. In grayscale images, the value of intensity at pixels may not be same but they must belong to a defined set of intensities.

Two pixels, p and q, are called m-adjacent (mixed adjacent) if these two pixels are either 4-adjacent or diagonal adjacent but not both. m-adjacency is an improvement over 8-adjacency as sometimes 8-adjacency may be ambiguous.

2.6 Digital Path

A (digital) path (or curve) from a pixel p (x_0, y_0) to pixel $q(x_n, y_n)$ is a sequence of distinct pixels with coordinates (x_0, y_0), (x_1, y_1), ..., (x_n, y_n), where (x_i, y_i) and (x_{i-1}, y_{i-1}) are adjacent for $1 \leq i \leq n$. If $(x_0, y_0) = (x_n, y_n)$, the path is known as a closed path and n is the length of the path. 4-, 8-, and m-paths can be defined based on the type of adjacency used.

2.7 Connected Set

Let S represents a subset of pixels in an image. Two pixels $p(x_0, y_0)$ and $q (x_n, y_n)$ are said to be connected in S if there exists a path (x_0, y_0), (x_1, y_1), ..., (x_n, y_n), where $\forall i, 0 \leq i \leq n, (x_i, y_i) \in S$. This set S is called a region of the image if the set S is a connected component and two regions are known to be adjacent to each other if their union forms a connected component, otherwise regions are said to be disjoint.

The boundary of the region S is the set of pixels in the region that have one or more neighbors that are not in S.

Summary

- An image is sensed and converted into digital form by the process of sampling and quantization.
- Sampling determines the spatial resolution of a digital image.
- Quantization determines the number of gray levels in the digital image.
- The quality of an image and file size is based on sampling and quantization.
- A color model is a well-defined system for generating a spectrum of colors.
- The two basic color models types are: Additive color model and Subtractive color model.
- Some common color models are RGB, CMY, CMYK and HSI.
- 4-Neigbours of a pixel are the pixels which are on the left, right, top and bottom of the pixel.
- 8-Neighbors of a pixel are the pixels which are on the left, right, top, bottom of the pixel and also diagonal neighbors of the pixel.

3
Mathematical Tools for Image Processing

3.1 Introduction

In various image processing activities, mathematical operations are very important. An image itself is represented in the form of a numerical matrix, having intensity or color information at each cell. In image processing operations, a large number of mathematical operations are used, e.g., interpolation in zooming and set operations in mathematical morphological operations. Various mathematical concepts, useful in implementing a number of image processing operations, are explained below.

3.2 Distance Function

Digital images follow certain metric and topological properties.

A distance function or metric D on a set S is a bivariate operator, i.e., it takes two operands, say $x \in S$ and $y \in S$, in order to map the set of non-negative real numbers, $[0, \infty)$.

$$D : S \times S \rightarrow [0,\infty)$$

A distance measure D is said to be a valid distance metric if, for all x, y, $z \in S$, the following conditions are satisfied:

$$
\begin{aligned}
&D(x,y) \geq 0 \; ; D(x, y) = 0 \text{ if } x = y \text{ (non-negative or separation axiom)} \\
&D(x, y) = D(y, x) \qquad\qquad\quad \text{(symmetry)} \\
&D(x, z) \leq D(x, y) + D(y, z) \quad \text{(triangle inequality or subadditivity)}
\end{aligned}
\tag{3.1}
$$

The distance between two pixels may be defined in various ways. A very simple way to define the distance is Euclidean distance, used in simple geometry.

Euclidean distance, also known as Pythagoras distance, is a simple straight-line distance defined as:

$$D((x_1,y_1), (x_2, y_2)) = \sqrt{((x_1 - x_2)^2 - (y_1 - y_2)^2)} \qquad (3.2)$$

Apart from Euclidean, another popular distance metric is Manhattan distance, also known as 'taxi-cab' distance or 'city block' distance, defined as:

$$D_4((x_1, y_1), (x_2, y_2)) = |x_1 - x_2| + |y_1 - y_2| \qquad (3.3)$$

City block distance D_4 is the minimum number of elementary steps required to move from a starting point to the end point in a grid, withonly horizontal and vertical moves being permitted. If diagonal moves are also permitted then another distance D_8, called Chessboard distance, is obtained. Chessboard distance, or Chebyshev distance, is defined as:

$$D_8((x_1, y_1), (x_2, y_2)) = \max\{|x_1 - x_2|, |y_1 - y_2|\} \qquad (3.4)$$

This distance computes the distance between two pixels by taking the maximum of their differences along any coordinate dimension. In two-dimensional space, this distance is equivalent to the minimum number of moves a king needs to make in order to travel between any two squares on the chessboard, hence the name Chessboard Distance.

Two pixels are called 4-neighbors if the D_4 distance between these two pixels is 1. This means that in 4-adjacency neighborhood only left, right, top and down adjacent pixels are considered while in 8-adjacency, the diagonal neighbors are also included. Two pixels are called 8-neighbors if the D_8 distance between these two pixels is 1.

If there exists a path between two pixels in the set of pixels in the image, then these pixels are known as contiguous pixels.

3.3 Convexity Property

If a line connecting any two points within a region lies within the region, then the region is defined as convex, otherwise it is non-convex.

3.4 Topological Properties

Topology or rubber sheet distortions is the study of properties of any object that remains unaffected by any deformation as long as there is no joining or tearing of the figure. For example, if any object has three holes in it, then even after rotating the object or stretching it, the number of holes will remain the same. Topological properties do not depend on the concept of distance. Any object that has a non-regular shape can be represented using its topological components.

3.5 Interpolation

Interpolation is the process used to find any missing value in a set of given values. For example, the value of a variable y is known at x = 1, 2, 4, 5, so the process used to find the value of y at x = 3 is known as interpolation. This is very useful in a number of image operations where there is a need to find the unknown value at any pixel, based on the known pixel values, i.e., where resampling is required, e.g., zooming of an image. A number of interpolation techniques are available in the literature. Traditional interpolating techniques are:

3.5.1 Nearest Neighbor Interpolation

This is a very simple interpolation technique, in which the nearest pixel intensity is replicated in order to find the unknown intensity value at any pixel location.

3.5.2 Bilinear Interpolation

In the bilinear interpolation technique, the weighted average of intensities of 2 x 2 neighboring pixels is used to interpolate an unknown intensity value.

In bilinear interpolation, the value of a pixel $x_{i,j}$ at the location (i,j) can be calculated as:

$$f(x) = a + \Delta i (b - a) + \Delta j (c - a) + \Delta i \Delta j (a - b - c + d) \tag{3.5}$$

where a, b, c, d are the nearest known pixel values to x; Δi and Δj define the relative distance from a to x (varying from 0 to 1).

3.5.3 Bicubic Interpolation

Bicubic interpolation uses the weighted average of the values of intensities of 4 x 4 neighboring pixels for finding the unknown intensity value at a pixel.

3.5.4 Cubic Spline Interpolation

Given a function f defined on [a, b] and a set of nodes $a = x_0 < x_1 < \cdots < x_n = b$, a cubic spline interpolant S (Fig. 3.1) for f is a function that satisfies the following conditions:

a) S(x) is a cubic polynomial, denoted $S_j(x)$, on the subinterval $[x_j, x_{j+1}]$, for each $j = 0, 1, \ldots, n-1$;

b) $S_j(x_j) = f(x_j)$ and $S_j(x_{j+1}) = f(x_{j+1})$ for each $j = 0, 1, \ldots, n-2$;

c) $S_{j+1}(x_{j+1}) = S_j(x_{j+1})$ for each $j = 0, 1, \ldots, n-2$;

d) $S'_{j+1}(x_{j+1}) = S'_j(x_{j+1})$ for each $j = 0, 1, \ldots, n-2$;

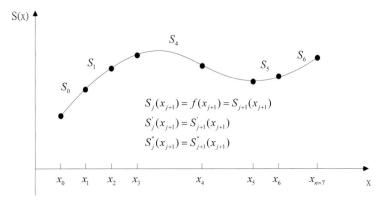

Fig. 3.1. Cubic Spline Interpolation.

e) $S''_{j+1}(x_{j+1}) = S''_j(x_{j+1})$ for each $j = 0, 1, \ldots, n-2$;

f) One of the following sets of boundary conditions is satisfied:

i) $S''(x_0) = S''(x_n) = 0$ (free or natural boundary);

ii) $S'(x_0) = f'(x_0)$ and $S'(x_n) = f'(x_n)$ (clamped boundary).

To construct the cubic spline interpolant for a given function f, the conditions in the definition are applied to the cubic polynomials

$$S_j(x) = a_j + b_j(x - x_j) + c_j(x - x_j)^2 + d_j(x - x_j)^3$$

for each $j = 0, 1, \ldots, n-1$. $\hspace{2cm}$ (3.6)

3.6 Circularly Symmetric Signals

An arbitrary 2D signal $a(x, y)$ can be expressed in the polar coordinate system as $a(r, \theta)$. When the 2D signal exhibits a circular symmetry,

$$a(x, y) = a(r, \theta) = a(r) \hspace{2cm} (3.7)$$

where $r^2 = x^2 + y^2$ and $\tan \theta = y/x$. A number of physical systems exhibit circular symmetry (e.g., lenses), it is useful to be able to compute an appropriate Fourier representation.

3.7 Statistics

- Probability distribution function of the brightness
- Probability density function of the brightness
- Average
- Standard deviation
- Coefficient-of-variation
- Signal-to-Noise (SNR) ratio

In image processing it is quite common to use simple statistical descriptions of images and sub-images. The notion of a statistics is intimately connected to the concept of a probability distribution, generally the distribution of signal amplitudes. For a given region, which could conceivably be an entire image, the probability distribution function of the brightness in that region and probability density function of the brightness in that region can be defined.

3.7.1 Probability Distribution Function

The probability distribution function $P(a)$ is the probability that a brightness chosen from the region is less than or equal to a given brightness value a. As a increases from $-\infty$ to $+\infty$, $P(a)$ increases from 0 to 1. $P(a)$ is monotonic, non-decreasing in a and thus $dP/da >= 0$.

3.7.2 Probability Density Function

The probability that a brightness in a region falls between a and $a + \Delta a$, given the probability distribution function $P(a)$, can be expressed as $p(a)\Delta a$ where $p(a)$ is the probability density function.

$$p(a)\Delta a = \left(\frac{dp(a)}{da}\right)\Delta a \tag{3.8}$$

Because of monotonic, non-decreasing character of $p(a)$ we have $P(a) \geq 0$ and $\int_{-\infty}^{\infty} p(a)da = 1$. For an image with quantized (integer) brightness amplitudes, the interpretation of Δa is the width of a brightness interval. The brightness probability density function is frequently estimated by counting the number of times that each brightness occurs in the region in order to generate a histogram, $h[a]$. The histogram can then be normalized so that the total area under the histogram is 1. Or the $p(a)$ for region is the normalized count of the number of pixels, N, in a region that have quantized brightness a:

$$p[a] = \frac{1}{N}h[a] \quad with \quad N = \sum_{\alpha} h[a] \tag{3.9}$$

Both the distribution function and the histogram as measured from a region are statistical descriptions of that region. It must be emphasized that both $P(a)$ and $p(a)$ should be viewed as estimates of true distributions when they are computed from a specific region.

3.7.3 Average

The average brightness of a region is defined as the sample mean of the pixel brightness within that region. The average, m_a, of the brightness over the N pixels within a region is given by:

$$m_\alpha = \frac{1}{N_{m,n \varepsilon R}} \sum a[m,n] \tag{3.10}$$

3.7.4 Standard Deviation

The unbiased estimate of the standard deviation, S_a, of the brightness within a region R with N pixels is called the sample standard deviation and is given by:

$$S_a = \sqrt{\frac{1}{N-1} \sum_{m,n \varepsilon R} (a[m,n] - m_a)^2}$$

$$= \sqrt{\frac{\sum\limits_{m,n \varepsilon R} a^2[m,n] - N m_a^2}{N-1}} \tag{3.11}$$

Using the histogram formulation gives:

$$S_a = \sqrt{\frac{\left(\sum\limits_a a^2 - h[a] - N \cdot m_a^2\right)}{N-1}}$$

The standard deviation, S_a, is an estimate σ_a of the underlying brightness probability distribution.

3.7.5 Coefficient-of-variation

The dimensionless coefficient-of-variation is defined as:

$$CV = \frac{S_a}{m_a} \times 100\% \tag{3.12}$$

3.7.6 Percentiles

The percentile, p%, of an unquantized brightness distribution is defined as the value of the brightness such that:

$$P(a) = p\%$$

or equivalently

$$\int_{-\infty}^{a} p(\alpha)d\alpha = p\%$$

Three special cases are frequently used in digital image processing.

- 0% of the minimum value in the region
- 50% of the median value in the region
- 100% of the maximum value in the region.

3.7.7 Mode

The mode of the distribution is the most frequent brightness value. There is no guarantee that a mode exists or that it is unique.

3.7.8 Signal-to-Noise Ratio

The noise is characterized by its standard deviation, S_n. The characterization of the signal can differ. If the signal is known to lie between two boundaries, $a_{min} <= a <= a_{max}$, then the signal-to-noise ratio (SNR) is defined as:

$$\text{Bounded signal } SNR = 20 \, log_{10} \left(\frac{S_a}{S_n} \right) \text{ dB} \tag{3.13}$$

3.8 Transforms

An image can be represented using pixel values (spatial domain) or using frequency spectrum (frequency domain). In frequency domain, an image can be expressed as a linear combination of some basis functions of some linear integral transform, e.g., Fourier transform, cosine transform, wavelet transform.

3.8.1 Two dimensional signals (images)

As a one-dimensional signal can be represented by an orthonormal set of basis vectors, an image can also be expanded in terms of a discrete set of basis arrays, called basis images, through a two dimensional (image) transform.

For an $N \times N$ image $f(x, y)$ the forward and inverse transforms are:

$$g(u,v) = \sum_{x=0}^{N-1} \sum_{y=0}^{N-1} T(u,v,x,y) f(x,y)$$

$$f(x,y) = \sum_{u=0}^{N-1} \sum_{v=0}^{N-1} I(x,y,u,v) g(u,v) \tag{3.14}$$

where, again, $T(u, v, x, y)$ and $I(x, y, u, v)$ are called the forward and inverse transformation kernels, respectively.

The forward kernel is said to be separable if

$$T(u, v, x, y) = T_1(u, x) \, T_2(v, y) \tag{3.15}$$

It is said to be symmetric if T_1 is functionally equal to T_2 such that

$$T(u,v,x,y) = T_1(u,x) T_1(v,y) \tag{3.16}$$

These properties are valid for the inverse kernel.

If the kernel $T(u, v, x, y)$ of an image transform is separable and symmetric, then

the transform $g(u,v) = \sum_{x=0}^{N-1}\sum_{y=0}^{N-1} T(u,v,x,y)f(x,y) = \sum_{x=0}^{N-1}\sum_{y=0}^{N-1} T_1(u,x)T_1(v,y)f(x,y)$ can

be written in matrix form:

$$g = T_1 \cdot f \cdot T_1^T \tag{3.17}$$

where f is the original image of size $N \times N$, and T_1 is an $N \times N$ transformation matrix with elements $t_{ij} = T_1(i, j)$. If, in addition, T_1 is a unitary matrix then the transform is called separable unitary and the original image is recovered through the relationship

$$f = T_1^{*T} \cdot g \cdot T_1^* \tag{3.18}$$

3.8.2 Discrete Fourier Transform

• Continuous space and continuous frequency

The Fourier transform is extended to a function $f(x, y)$ of two variables. If $f(x, y)$ is continuous and integrable and $F(u, v)$ is integrable, the following Fourier transform pair exists:

$$F(u,v) = \int\int_{-\infty}^{\infty} f(x,y)e^{-j2\pi(ux+vy)}dxdy \tag{3.19}$$

$$f(x,y) = \frac{1}{(2\pi)^2}\int\int_{-\infty}^{\infty} F(u,v)e^{j2\pi(ux+vy)}dudv \tag{3.20}$$

In general, $F(u, v)$ is a complex-valued function of two real frequency variables u, v and hence, it can be written as:

$$F(u,v) = R(u,v) + jI(u,v) \tag{3.21}$$

The amplitude spectrum, phase spectrum and power spectrum, respectively, are defined as follows:

$$|F(u,v)| = \sqrt{R^2(u,v) + I^2(u,v)}$$

$$\phi(u,v) = \tan^{-1}\left[\frac{I(u,v)}{R(u,v)}\right]$$

$$P(u,v) = |F(u,v)|^2 = R^2(u,v) + I^2(u,v) \tag{3.22}$$

• Discrete space and continuous frequency

For the case of a discrete sequence $f(x, y)$ of infinite duration, we can define the 2-D discrete space Fourier transform pair as follows:

$$F(u,v) = \sum_{x=-\infty}^{\infty} \sum_{y=-\infty}^{\infty} f(x,y)e^{-j(xu+vy)} \tag{3.23}$$

$$f(x,y) = \frac{1}{(2\pi)^2} \int_{u=-\pi}^{\pi} \int_{v=-\pi}^{\pi} F(u,v)e^{j(xu+vy)} \, du \, dv \tag{3.24}$$

$F(u, v)$ is, again, a complex-valued function of two real frequency variables u, v and it is periodic with a period of $2\pi \times 2\pi$, that is $F(u,v) = F(u+2\pi,v) = F(u,v+2\pi)$.

The Fourier transform of $f(x, y)$ is said to converge uniformly when $F(u, v)$ is finite and

$$\lim_{N_1 \to \infty} \lim_{N_2 \to \infty} \sum_{x=-N_1}^{N_1} \sum_{y=-N_2}^{N_2} f(x,y)e^{-j(xu+vy)} = F(u,v) \text{ for all } u, v. \tag{3.25}$$

when the Fourier transform of $f(x, y)$ converges uniformly, $F(u, v)$ is an analytic function and is infinitely differentiable with respect to u and v.

• **Discrete space and discrete frequency: The two-dimensional Discrete Fourier Transform (2-D DFT)**

If $f(x, y)$ is an $M \times N$ array, such as the one obtained by sampling a continuous function of two dimensions at dimensions M and N on a rectangular grid, then its two dimensional Discrete Fourier transform (DFT) is the array given by

$$F(u,v) = \frac{1}{MN} \sum_{x=0}^{M-1} \sum_{y=0}^{N-1} f(x,y)e^{-j2\pi(ux/M+vy/N)} \tag{3.26}$$

where u = 0, 1, 2, ...M–1 and v = 0, 1 ,2 .... N–1

and the inverse DFT (IDFT) is

$$f(x,y) = \sum_{u=0}^{M-1} \sum_{v=0}^{N-1} F(u,v)e^{j2\pi(ux/M+vy/N)} \tag{3.27}$$

where x = 0, 1,2, ... M–1 and y = 0, 1, 2, ... N–1

When images are sampled in a square array, i.e., $M = N$, then

$$F(u,v) = \frac{1}{N} \sum_{x=0}^{N-1} \sum_{y=0}^{N-1} f(x,y)e^{-j2\pi(ux+vy)/N} \tag{3.28}$$

$$f(x,y) = \frac{1}{N} \sum_{u=0}^{N-1} \sum_{v=0}^{N-1} F(u,v)e^{j2\pi(ux+vy)/N} \tag{3.29}$$

Two-dimensional Discrete Fourier Transform is separable, symmetric and unitary.

3.9 Wavelet Transform

The decomposition of an image g into wavelet coefficients through a Discrete Wavelet Transform (DWT) W can be expressed as:

$$G = W(g) \tag{3.30}$$

Low-pass (approximation) and high-pass (detail) coefficients at each scale are obtained using:

$$W_\varphi(j_0, r, s) = \frac{1}{\sqrt{MN}} \sum_0^{r-1} \sum_0^{s-1} g(x, y)\, \varphi_{j_0, r, s}(x, y) \tag{3.31}$$

$$W_\psi^i(j, r, s) = \frac{1}{\sqrt{MN}} \sum_0^{r-1} \sum_0^{s-1} g(x, y)\, \psi_{j, r, s}^i(x, y) \tag{3.32}$$

$i = \{H, V, D\}$

where W_φ and W_ψ represent the approximation and detail (wavelet) coefficients, φ and ψ are the basis functions, index i identifies the horizontal, vertical and diagonal details, j represents the scale, j_0 is an arbitrary starting scale and r, s are the position-related parameters.

The successive applications of Eqs. (3.25) and (3.26) produce different frequency sub-bands labelled as LL_j, LH_j, HL_j and HH_j, $j = 1, 2, \ldots, J$, where the subscript indicates the j-th resolution level of wavelet transform and J is the largest scale in the decomposition. These sub-bands contain different information about the image. The lowest frequency LL_J sub-band corresponds to a rough approximation of the image signal, while the LH_j, HL_j and HH_j sub-bands correspond to horizontal, vertical and diagonal details of the image signal, respectively. The highest frequency HH_1 sub-band is the noisiest sub-band among all the sub-bands obtained at each resolution level. The LL_{j-1} sub-band can be recursively decomposed further in order to form the LH_j, HL_j and HH_j sub-bands. Figure 3.2 illustrates the sub-band regions of a critically sampled wavelet transform.

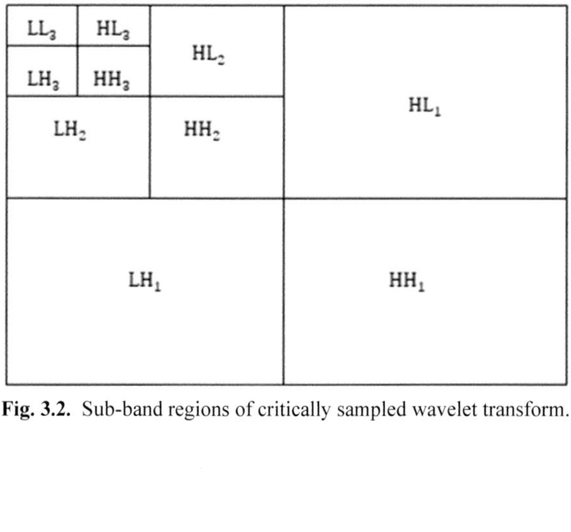

Fig. 3.2. Sub-band regions of critically sampled wavelet transform.

3.10 Discrete Cosine Transform (DCT)

For 2-D signals, DCT is defined as

$$C(u,v) = a(u)a(v)\sum_{x=0}^{N-1}\sum_{y=0}^{N-1} f(x,y)\cos\left[\frac{(2x+1)u\pi}{2N}\right]\cos\left[\frac{(2y+1)v\pi}{2N}\right] \tag{3.33}$$

$$f(x,y) = \sum_{u=0}^{N-1}\sum_{v=0}^{N-1} a(u)a(v)C(u,v)\cos\left[\frac{(2x+1)u\pi}{2N}\right]\cos\left[\frac{(2y+1)v\pi}{2N}\right] \tag{3.34}$$

$a(u)$ is defined as above and $u,\ v = 0,1\ldots,\ N-1$.

The DCT is a real transform and, therefore, better than the Fourier transform.

DCT also has excellent energy compaction properties and, therefore, is widely used in image compression standards (JPEG standards, for example).

3.11 Walsh Transform (WT)

The Walsh transform for two dimensional signals is defined as:

$$W(u,v) = \frac{1}{N}\sum_{x=0}^{N-1}\sum_{y=0}^{N-1} f(x,y)\left[\prod_{i=0}^{n-1}(-1)^{(b_i(x)b_{n-1-i}(u)+b_i(y)b_{n-1-i}(v))}\right] \tag{3.35}$$

or

$$W(u,v) = \frac{1}{N}\sum_{x=0}^{N-1}\sum_{y=0}^{N-1} f(x,y)(-1)^{\sum_{i=0}^{n-1}(b_i(x)b_{n-1-i}(u)+b_i(y)b_{n-1-i}(v))} \tag{3.36}$$

The inverse Walsh transform for two dimensional signals is defined as:

$$f(x,y) = \frac{1}{N}\sum_{u=0}^{N-1}\sum_{v=0}^{N-1} W(u,v)\left[\prod_{i=0}^{n-1}(-1)^{(b_i(x)b_{n-1-i}(u)+b_i(y)b_{n-1-i}(v))}\right] \tag{3.37}$$

or

$$f(x,y) = \frac{1}{N}\sum_{u=0}^{N-1}\sum_{v=0}^{N-1} W(u,v)(-1)^{\sum_{i=0}^{n-1}(b_i(x)b_{n-1-i}(u)+b_i(y)b_{n-1-i}(v))} \tag{3.38}$$

Unlike the Fourier transform, which is based on trigonometric terms, the Walsh transform consists of a series expansion of basis functions whose values are of −1 or 1 and whose form is of square waves. These functions can be implemented more efficiently in a digital environment than the exponential basis functions of the Fourier transform.

The forward and inverse Walsh kernels are identical, except for a constant multiplicative factor of $\frac{1}{N}$ for 1-D signals. The forward and inverse Walsh kernels are identical for 2-D signals. This is because the array formed by the kernels is a symmetric matrix with orthogonal rows and columns, so its inverse array is the same as the array itself.

3.12 Matrix Operations

Matrix operations are very important in the context of image processing as, mathematically, an image is represented in the form of a matrix. Various matrix operations are applied in image processing. Some common matrix operations are:

3.12.1 Matrix Addition

The sum of two matrices $A = [a_{ij}]_{mxn}$ and $B = [b_{ij}]_{mxn}$ is defined as:

$$A + B = [a_{ij} + b_{ij}]_{mxn} \qquad (3.39)$$

3.12.2 Matrix Subtraction

The difference of two matrices $A = [a_{ij}]_{mxn}$ and $B = [b_{ij}]_{mxn}$ is defined as:

$$A - B = [a_{ij} - b_{ij}]_{mxn} \qquad (3.40)$$

3.12.3 Matrix Multiplication

Multiplication of two matrices $A = [a_{ij}]_{mxn}$ and $B = [b_{jk}]_{nxp}$ is defined as:

$$A * B = \sum_{l=1..n} a_{il} \, b_{lk} \qquad (3.41)$$

It is important to note that the number of columns in the first matrix should be the same as the number of rows in second matrix. The resultant matrix is of dimension mxp. Another important point is that it is not necessary that $A * B \neq B * A$

3.12.4 Identity Matrix

A square matrix I is called an identity matrix or unit matrix if all the main diagonal elements are 1 and all other elements are 0. The result of the multiplication of a matrix by an Identity matrix is the same as the original matrix, i.e., $A*I = I*A = A$.

3.12.5 Inverse Matrix

A matrix A^{-1} is said to be inverse of matrix A, if $A * A^{-1} = A^{-1} * A = I$ (Identity matrix)

3.13 Set Theory

The concept of set theory is very useful in image processing, particularly in mathematical morphology. In morphology, an image is considered as a set and various set operations are performed on the image.

A set is defined as a well-defined collection of objects. The objects are called the elements.

- If every element of a set A is also an element of a set B, then we say that A is a sub-set of B (A $\subseteq$ B).
- If A and B are sets, such that A $\subseteq$ B but A $\neq$ B, then A is a proper sub-set of B (A $\subset$ B).
- Two sets A and B are equal if they have exactly the same elements.
- The universal set is the set of all elements of interest in a particular application. It is the largest, in the sense that all sets considered in that particular application are sub-sets of the universal set.
- The union of sets A and B (A $\cup$ B) is the set of all elements that belong to either A or B or both. A $\cup$ B = {x | x $\in$ A or x $\in$ B or both}
- The intersection of sets A and B (A $\cap$ B) is the set of all elements in common with the sets A and B. A $\cap$ B = {x | x $\in$ A and x $\in$ B}
- Two sets are said to be disjoint if have no elements in common, i.e., if A $\cap$ B = $\emptyset$.
- If U is a universal set and is a sub-set of U, then the set of all elements in U that are not in A is called the complement of A and is denoted A^C.

$$A^C = \{x \mid x \in U, x \notin A\}$$

Summary

- A digital image is mathematically represented in the form of a matrix, therefore mathematical operations can easily be performed on the digital images.
- Various distance functions are available that are used in various digital image processing operations, e.g., Euclidean distance, D_4 Distance, D_8 distance.
- Topology is the study of the properties of any object that remains unaffected by any deformation, as long as there is no joining or tearing of the figure.
- Mathematical transforms are used to convert the image from one transform to another.
- A number of transforms are used in image processing, such as Discrete Sine Transform, Discrete Cosine Transform, Fourier Transform, Hough Transform and Wavelet Transform.
- Matrix operations addition, subtraction, multiplication, etc., can be applied on digital images in order to perform various image processing actions.
- An image can be represented as a set of pixels and set operations can be applied on the images.
- Mathematical morphology uses set operations for image processing.

4

Image Enhancement in Spatial Domain

4.1 Introduction

Image enhancement is the process of editing an image in order to make it 'better' for a human viewer for a particular application. This involves smoothing or sharpening image contents. Spatial domain refers to the two-dimensional image plane in terms of pixel intensities. In spatial domain image enhancement methods, the image is processed by operating on the pixel intensities directly in the image plane itself. The other category of image enhancement methods is transform domain methods. Transform domain methods operate on the Fourier transform of an image.

Spatial domain transformations are of the form:

$$g(x,y) = T[f(x,y)], \tag{4.1}$$

where $f(x,y)$ is the input image, T is transformation to be applied on the image and $g(x,y)$ is the transformed image

A number of spatial domain methods for image enhancement are available. In the next part of this chapter, the main spatial domain enhancement methods are described.

4.2 Point Processing

In point processing, each pixel is operated independently. Only the intensity of the pixel being processed is considered. In these operations, the neighboring pixels are not considered in the processing, i.e., the neighborhood size is 1x1, which is smallest possible. In point processing, every pixel with the same graylevel value in the input image maps to the same value in the output image. Point operations require very little memory.

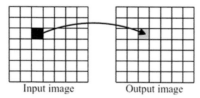

Input image   Output image

Fig. 4.1 Point processing.

4.2.1 Brightness Manipulation

Point processing is very useful in the brightness manipulation of an image. The brightness is associated with each pixel in an image. To increase or decrease the brightness of an image, a constant value can be added or subtracted from each and every pixel of the image.

To increase the brightness of an image $f(x,y)$, the operation is:

$$g(x,y) = f(x,y) + c \qquad (4.2)$$

To decrease the brightness of an image $f(x,y)$, the operation is:

$$g(x,y) = f(x,y) - c \qquad (4.3)$$

where c is a constant and represents the value to be added or subtracted to change the brightness of an image.

This operation can be performed in MATLAB®, using functions similar to:

```
a = imread('filename');
b = a + 30; (to increase the brightness by a factor of 30)
b = a – 30; (to decrease the brightness by a factor of 30)
figure, imshow (b);
```

The effect of these functions on an image is shown in Fig. 4.2.

4.2.2 Inverse Transformation

This transformation is also known as Image negative. This transformation reverses the intensity levels of an image at all pixels. Dark is converted into bright and bright is converted into dark. This transformation is obtained by subtracting the value at each pixel from the maximum pixel value and is represented as:

$$g(x,y) = \text{Max. pixel value} - f(x,y) \qquad (4.4)$$

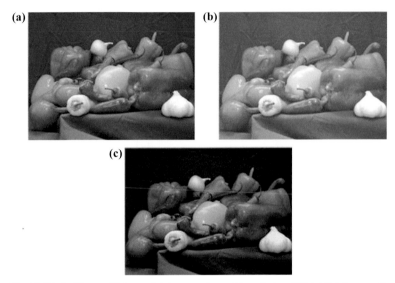

Fig. 4.2. (a) Original image (b) Image brightness enhanced by a factor of 30 (c) Brightness of image (a) reduced by a factor of 30.
Original image reprinted with permission of The MathWorks, Inc.
(For color image of Fig. 4.2(a), see Fig. 1.1 (a) in Color Figures Section at the end of the book)

Inverse transformation is useful in those applications where there is a need to highlight image details embedded into dark regions of an image, in particular when the black areas are dominant in size.

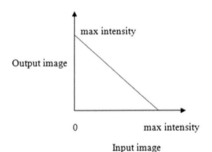

Fig. 4.3. Negative transformation.

In MATLAB, an image negative can be obtained using a function sequence for a maximum image intensity value of 256:

a = imread('filename');
b = 256-a;
figure, imshow (b);

Figure 4.4 shows an original image and its corresponding negative image using the above function sequence;

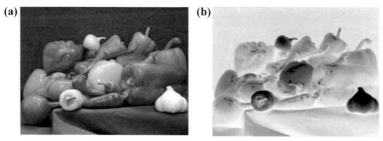

Fig. 4.4. (a) Original image (b) Corresponding Negative image.
Original image reprinted with permission of The MathWorks, Inc.
(For color image of Fig. 4.4(a), see Fig. 1.1 (a) in Color Figures Section at the end of the book)

4.2.3 *Contrast Manipulation*

Point processing operations can also be used to manipulate the contrast information of an image. In this process, the range of pixel luminance values is changed.

Contrast manipulation of an image f(x,y) can be represented as:

g(x,y) = f(x,y)*c, where c is the contrast manipulation factor.   (4.5)

A value greater than 1 increases the contrast and expands the range of luminance values. As a result, the bright values become brighter and dark values become darker. The values less than 1 reduce the range of luminance.

Fig. 4.5. (a) An image (b) Corresponding grayscale image (c) and (d) Contrast enhancement applied with factor .20 and 5, respectively.
(For color image of Figure 4.5 (a) see Color Figures Section at the end of the book)

In MATLAB, it can be done using a function sequence:

```
a=imread(filename);
b=rgb2gray(a);
c=b*.20;
figure, imshow(c);
d=b*5;
figure, imshow(d)
```

The effect of these functions is shown in Fig. 4.5.

4.2.4 Log Transformations

These transformations are non-linear graylevel transformations and are of the form:

$$g(x,y) = c \log (1+ f(x,y)). \tag{4.6}$$

where c is a constant.

This transformation increases the range of graylevel values in the output image. The lower greylevel values are spread out. The input image, having a narrow range of low intensity values, is converted into an image that has a wider range of output levels. To achieve the opposite effect, an inverse log transformation can be used.

4.2.5 Power-law Transformations

Power-law transformation is given by:

$$g(x,y) = c \, [f(x,y)]^{\gamma} \tag{4.7}$$

where c and γ are positive constants.

Power-law transformations are also known as gamma transformation, due to exponent γ. By changing the values of γ, a family of transformations is obtained. When the value of γ is less than 1, the image appears to be dark and when it is greater than 1, the image appears bright. If $c = \gamma = 1$, the power-law transformation is converted into an identity transformation.

Many devices, e.g., a monitor used for input and output, respond according to a power-law. The process used in order to reduce the effect of power-law response phenomena is called gamma correction. For example, on some monitors, the image displayed is darker than the actual input image. In such cases, there is a need to correct this effect before displaying the image. Gamma correction changes the intensity of the displayed image as well as the ratios of red to green to blue in a color image. Gamma correction is very important in the case of the images used on websites, as the image has to be displayed on various monitors having different values of gamma.

4.2.6 Piecewise-Linear Transformations

Piecewise-linear transformations are those non-linear transformations which are applied to input image pixels based on some criteria. For the same image pixels, the transformation applied can be different. There are various piecewise linear image transformations. Some are:

- **Intensity-level slicing**

    To highlight a specific range of gray values, intensity level slicing can be used. If in any application there is a need to emphasize on a set of a particular range of intensities, instead of the all values, then intensity-level slicing is very useful. Although intensity-level slicing can be applied in various ways, the most frequently used ways are (i) Highlighting the intensities in a defined range [A,B] and reducing all other intensities to a lower level (ii) Highlighting the intensities in a defined range [A,B] and preserving all other intensities.

- **Thresholding**

    A thresholding transformation is used to extract a part of an image based on an intensity value. In this transformation a threshold value has to be defined. The pixels having intensity greater than the defined threshold value are set to a particular value and others are left unchanged, or vice-versa.

$$g(x, y) = \begin{cases} 0 & ; if(x, y) > threshold\ value \\ 255 & ; otherwise \end{cases} \qquad (4.8)$$

where f(x,y) is the input image and g(x,y) is the transformed image.

4.3 Mask Processing

In spatial domain enhancement methods, if neighbors are also considered in processing the value of the pixel, the processing is called Mask Processing. The pixel, being processed along with the neighbors considered, defines a mask. This mask is moved from pixel to pixel over the whole image in order to generate the processed image. This mask in spatial domain is also called window, kernel or template. In general, the mask is taken of size nxn, where n is an odd number. The center pixel of the mask, called the origin, coincides with the pixel being operated at a particular instance, while processing. For a non-symmetric mask, any pixel of the mask can be considered as the origin, depending on the use.

The processed value is calculated as follows:

$g(x,y) = w_{-1-1}*f(x-1, y-1) + w_{-10}*f(x-1,y) + w_{-11}*f(x-1,y+1)$
$+w_{0-1}*f(x, y-1)+ w_{00}*f(x,y) +w_{01}*f(x,y+1) + w_{1-1}*f(x+1,y-1) + w_{10}*f(x+1,y) +$
$w_{11}*f(x+1,y+1)$ \hfill (4.9)

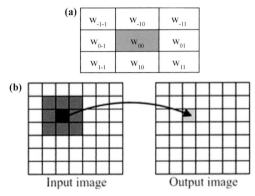

Fig. 4.6. (a) A mask of size 3x3. (b) During processing the dark shaded pixel of the mask coincides with the pixel being processed.

where g(x,y) is the processed value of a pixel f(x,y) using a mask as shown in Fig. 4.6 and Fig. 4.7. w denotes the weight assigned to the pixel value at neighboring pixels.

In general, spatial filtering can be defined as:

$$g(x, y) = \sum_{s=-a}^{a} \sum_{t=-b}^{b} w(s, t) f(x + s, y + t)$$ (4.10)

where for a mask of size mxn, m = 2a + 1 and n = 2b + 1, a and b are positive integers. x and y are varied so that each pixel in w visits every pixel in f.

Mask processing filters can be divided into two categories: Smoothing filters (Low pass filters) and Sharpening filters (High pass filters).

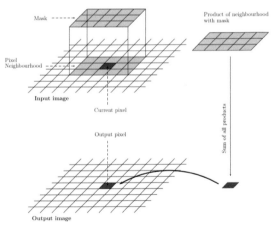

Fig. 4.7: Mask processing.

4.4 Smoothing Filters

Smoothing filters remove small details of an image and bridge small gaps in lines or curves. Smoothing is used for blurring and for noise reduction in an image. Smoothing filters are also known as averaging filters, because these filters replace the pixel intensity by the average of intensity values of its neighbors.

4.4.1 Smoothing Linear Filters

Smoothing linear filters are also called averaging filters. In these filters, every pixel value is replaced by the average value of intensities values of its neighborhood. A simple smoothing filter mask is box filter

A spatial averaging filter with all the same coefficients is called box filter. In a box filter, equal weight is assigned to all pixel intensities (Fig. 4.8). If there is a need to assign more importance to some pixels in comparison to others, then the weights in the filter can be changed accordingly. The following smoothing filter (Fig. 4.9) assigns more weight to the center pixel in comparison to its other neighbors. And among neighbors, a higher weight is assigned to top, down, left and right neighbors than to diagonal neighbors.

Figure 4.10 shows the effect of smoothing by box filters of various sizes. The effect of averaging or smoothing is clearly visible in the figures. The blurring effect increases as the size of the mask increases. Therefore, the size of the mask has to be selected very carefully depending on the application.

$(1/16)$ x

1	2	1
2	4	2
1	2	1

Fig. 4.8. A smoothing box filter of size 3x3.

$(1/9)$ x

1	1	1
1	1	1
1	1	1

Fig. 4.9. A weighted average filter of size 3x3.

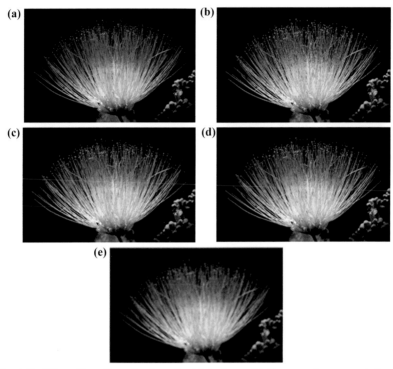

Fig. 4.10. Effect of box filter of various size (a) An image (b) Corresponding grayscale image (c) Image processed with box filter of size 3x3 (d) Image processed with box filter of size 5x5 (e) Image processed with box filter of size 9x9.
(For color image of Fig. 4.10 (a) see Color Figures Section at the end of the book)

MATLAB function for this is imboxfilt. The size of the mask can be described in the function. The MATLAB code is of the form:

```
clc
clear all
close all
a=imread('flowers.jpeg');
 b=rgb2gray(a);
c=imboxfilt(b,9);
figure, imshow(c)
```

4.4.2 Smoothing Non-Linear Filters

In smoothing non-linear filters, the intensity value of a pixel is replaced by the value of the pixel of the filter area selected by some ranking. A very common non-linear smoothing filter is the median filter, in which the value of a pixel is replaced by the median intensity value of all pixels' intensity in the neighborhood of the pixel. To implement a median filter, initially all the values on the defined neighborhood, including the pixel being processed, are sorted. The median value is selected, and the pixel being processed is assigned this median intensity value. Median filters are very useful in removing salt-and-pepper noise, in which the intensity of certain pixels is very different from the neighboring pixels. By applying median filters, these noisy isolated pixels are replaced by the value more like the neighbors, as extreme values are not considered in this filter. In a 3x3 neighborhood, the 5th largest value is the median and that value will be used for replacement at the pixel being processed. Similarly, in a 5x4 neighborhood, 13th highest value will be considered. In the case of median filters, isolated pixel clusters which are very bright or very dark in comparison to their neighbors and cover an area of less than $m^2/2$ are removed by an mxm size median filter.

Other filters in the order-statistics filter category are max filter and min filter. These filters select the maximum intensity or minimum intensity of all pixels in the defined neighborhood. These filters are very useful in those applications, where the brightest or darkest pixels in the image are required to be filtered.

Generally, smoothing filters are used in conjunction with thresholding in those applications. Initially, the objects are blurred using smoothing filters and then, by thresholding, the less visible or blurred objects are filtered out.

- **Gaussian low pass filter:** A Gaussian low pass filter is described as:

$$G(x, y) = \frac{1}{2\pi\sigma^2} e^{-\frac{x^2+y^2}{2\sigma^2}}, \text{ where } \sigma \text{ is standard deviation of the distribution.}$$

The effect of Gaussian low pass filters with various standard deviations is shown in Fig. 4.11.

4.5 Sharpening Filters

Sharpening is opposite to smoothing process. Sharpening filters are used to sharpen the objects in an image, i.e., to differentiate between the objects or background. Mathematically, this process is spatial differentiation. This process is very useful in those applications where fine details are required to be observed in an image, e.g., medical applications where every small deformity in the image needs to be observed. An image sharpening filter is shown in Fig. 4.12.

Figure 4.13 shows the effect of Smoothing, Sharpening and Median filters on the numerical values of an effect. The change in numerical values (in box, after processing) clearly indicates the effect of various filters. It can be observed that in smoothing, the value of the pixels, as well as the value of neighboring pixels, is changed so that the difference in intensities is reduced. In the case of the sharpening filter, the difference in intensities is increased and so the difference in intensity values is also increased. In the case of median filtering, the values of pixels having different intensity are changed so as to match with neighboring pixels, without affecting the intensity values of neighboring pixels.

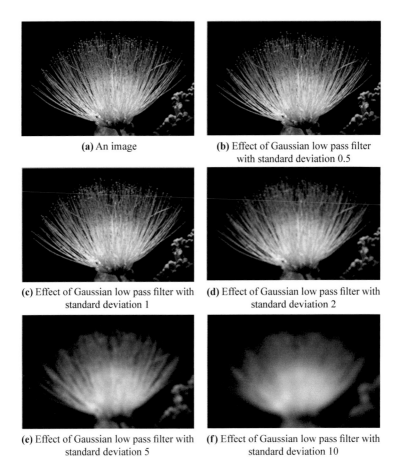

(a) An image

(b) Effect of Gaussian low pass filter with standard deviation 0.5

(c) Effect of Gaussian low pass filter with standard deviation 1

(d) Effect of Gaussian low pass filter with standard deviation 2

(e) Effect of Gaussian low pass filter with standard deviation 5

(f) Effect of Gaussian low pass filter with standard deviation 10

Fig. 4.11. An image and effect of Gaussian low pass filters with different standard deviation.

4.5.1 Unsharp Masking

Unsharp masking is a technique in which a blurred image is subtracted from the original image in order to get a sharper image. This process is adopted from the publishing industry. The process of unsharp masking is represented as:

$$\hat{f}(x, y) = f(x, y) + \propto [f(x, y) - \bar{f}(x, y)],$$

where $f(x, y)$ is the original image and $\bar{f}(x, y)$ is the corresponding blurred image, $\propto$ is the weight function   (4.11)

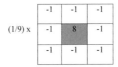

Fig. 4.12. An image sharpening filter.

Fig. 4.13. Effect of Smoothing, Sharpening and Median filters on an image in terms of numerical values.

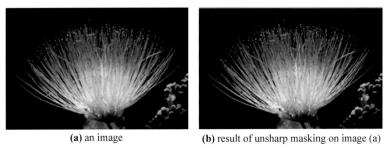

(a) an image   (b) result of unsharp masking on image (a)

Fig. 4.14. Unsharp masking.

in MATLAB, imsharpen function is used for image sharpening with unsharp masking.

The result of unsharp masking using MATLAB imsharpen function is shown in Fig. 4.14.

4.6 Bit-Plane Slicing

An image can be divided into various bit-planes, depending on the way intensities are stored. For example, an eight-bit image can be divided into eight bit-planes. Then processing can be done on each bit-plane independently. The change in the left-most bit, i.e., Most Significant Bit (MSB), significantly impacts the image, while the right-most bit, known as the Least Significant Bit (LSB), has very little visible impact on the image. Figure 4.15 shows an image and its various bit-planes separately. Figure 4.15j is a reconstructed image after adding these 8-bit planes, which is exactly same as Fig. 4.15a. The image in Fig. 4.15k is obtained by adding higher bits, which is visually similar to images 4.15a and 4.15j, but mathematically this image is only a 4-bit image. 4 low-bitplanes have been removed, having very little impact on visual quality. This concept is very useful in image compression techniques as the file size has been reduced to half without compromising the quality.

In MATLAB, the bitget function is used to fetch a bit from a specified position from all the pixels. For example bitget(A,8) returns the higher order bits of all pixels from an image A.

4.7 Arithmetic Operations

Various simple arithmatic operations, such as addition, subtraction and multiplication, can be performed on the images.

4.7.1 Image Addition

Using simple addition,two images f(x,y) and g(x,y) can be combined in order to get a new image h(x,y):

$$h(x,y) = f(x,y)+g(x,y) \tag{4.12}$$

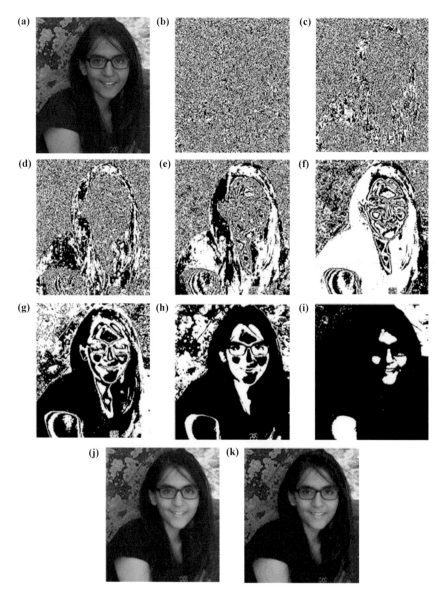

Fig. 4.15. (a) An 8-bit image (b)–(i) image divided into various bit planes, first bit (LSB) to 8th bit (MSB) (j) Image reconstructed by adding images (b) to (i), (k) image reconstructed by adding images (f) to (i).

The dimensions of the two images should be the same for this binary operation.

It can be done in MATLAB by simply using '+' operator, because MATLAB operates on the images in matrix form and image addition is addition of two matrices. Figure 4.16 shows the effect of adding two images together.

Fig. 4.16. Two images (first row) and addition of these two images (second row).

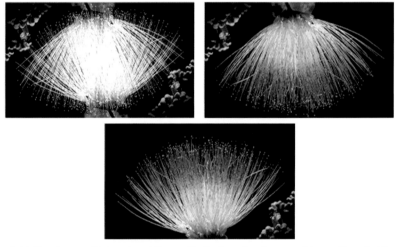

Fig. 4.17. Two images (first row) and the result of subtracting the second image from first image (second row).

4.7.2 Image Subtraction

The difference in two images f(x, y) and g(x, y) can be found by image subtraction, that can be performed as:

$$h(x, y) = f(x, y)-g(x, y) \tag{4.13}$$

Image g(x, y) will be deleted from image h(x, y) using this operation. The dimensions of the two images should be same for this binary operation. It may be noted that f–g ≠ g–f. The result will be different for both operations.

It can be done in MATLAB by simply using '–' operator, because MATLAB operates on the images in matrix form and image subtraction is subtraction of one matrix from another. Figure 4.17 shows the effect of subtraction of two images.

4.7.3 Image Multiplication

Image multiplication is used to extract a region of interest from an image. If a user is interested in a specific part of an image, then it can be isolated by multiplying the region of interest by 1 and the remaining part by 0. As a result, only the region of interest will contain image information and the remaining part will become 0 or black.

It can be done in MATLAB using function immultiply(X,Y). Using this function, either two images can be multiplied, or a constant can be multiplied to all values of the pixel values of an image.

4.7.4 Image Division

Image division is opposite to image multiplication. In MATLAB, image division can be performed using function imdivide (X, Y). Image division can be performed either by another image or a by a constant value.

4.8 Logical Operations

Logical operations AND, OR and NOT can be applied on binary images. AND and OR operations are binary operations, so two images are required in order to perform these operations, while the NOT operation is a unary operation and, hence, can be performed on a single image. These logical operations are similar to set operations. The AND operation corresponds to Intersection, the OR operation corresponds to Union and the NOT operation corresponds to Complement set operation.

If A and B are two sets, then logical operations are defined as:

A AND B = set of all elements which are common to set A and set B

A OR B = set of all elements which are either in set A or in set B

NOT A = set of all elements which are not in A, this operation is similar to the Image negative operation discussed earlier.

These are basic logical operations; other logical operations can be performed by combining these logical operations.

An important point to note here is that logical operations can be performed on images or objects of any size, while operations discussed in previous sections need both images to be of the same size.

4.9 Geometric Operations

There are a number of geometric transformations that can be applied to a digital image. These transformations are unary transformations, i.e., these transformations are applied on a single image. These transformations are:

4.9.1 Zooming and Shrinking

Zooming and Shrinking operations are used to resize an image. Resizing of an image can be done temporarily, for viewing, and also permanently. Resizing of an image, specifically for viewing purpose, is called as Zooming (in) and Shrinking (out), while resizing is changing an image in terms of value stored for long term use in the form of a file. Temporarily zooming and shrinking can be done interactively with the help of various tools available.

MATLAB uses the imresize function for resizing images. The imresize function allows the user to specify the interpolation method used (nearest-neighbor, bilinear, or bicubic interpolation).

4.9.2 Translation

Translation of an input image $f(x, y)$ with respect to its Cartesian origin, producing an output image $g(\dot{x}, \dot{y})$, is defined as:

$g(\dot{x}, \dot{y}) = f(x + \Delta x, y + \Delta y)$, where Δx and Δy are displacements in x and y directions, respectively. Translation is an affine transformation, i.e., it preserves straight lines and planes.

MATLAB uses the imtransform function for translation of an image.

4.9.3 Rotation

Rotation of an image is a spatial transformation in which an image is rotated at a specified angle. Rotation is also an affine geometric transformation.

4.9.4 Cropping

An image can be cropped to a specified angle, i.e., the part of an image can be cut.

In MATLAB, the imcrop tool facilitates cropping of an image to a specified rectangle interactively.

4.10 Image Padding

When performing processing on an image using a mask, it is difficult to process the image boundary pixels, as shown in Fig. 4.18, as there are no corresponding image pixels for some values of the mask. One solution to this problem is to skip

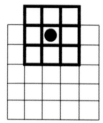

Fig. 4.18. Problem in mask processing at boundary pixels.

1	1	2	5	6	3	6	7	3
2	3	4	6	7	5	1	8	4
8	7	6	5	7	6	3	3	4
2	3	5	6	7	8	2	7	3
4	5	3	2	1	6	8	7	2
1	4	5	3	2	6	7	8	1
2	3	4	5	6	8	9	2	1

0	0	0	0	0	0	0	0	0	0	0	0	0
0	0	0	0	0	0	0	0	0	0	0	0	0
0	0	1	1	2	5	6	3	6	7	3	0	0
0	0	2	3	4	6	7	5	1	8	4	0	0
0	0	8	7	6	5	7	6	3	3	4	0	0
0	0	2	3	5	6	7	8	2	7	3	0	0
0	0	4	5	3	2	1	6	8	7	2	0	0
0	0	1	4	5	3	2	6	7	8	1	0	0
0	0	2	3	4	5	6	8	9	2	1	0	0
0	0	0	0	0	0	0	0	0	0	0	0	0
0	0	0	0	0	0	0	0	0	0	0	0	0

2	2	2	3	4	6	7	5	1	8	4	4	4
1	1	1	1	2	5	6	3	6	7	3	3	3
1	1	1	1	2	5	6	3	6	7	3	3	7
2	2	2	3	4	6	7	5	1	8	4	4	8
7	8	8	7	6	5	7	6	3	3	4	4	3
3	2	2	3	5	6	7	8	2	7	3	3	7
5	4	4	5	3	2	1	6	8	7	2	2	7
4	1	1	4	5	3	2	6	7	8	1	1	8
3	2	2	3	4	5	6	8	9	2	1	1	2
3	2	2	3	4	5	6	8	9	2	1	1	2
3	1	1	4	5	3	2	6	7	8	1	1	2

Fig. 4.19. An input image and corresponding 0-padded image and border replicated image.

the boundary pixels while processing and process only those pixels in an image which are completely overlapping with the mask. In this, some pixels remain unprocessed. If there is a need to process all the pixels, image padding can be used. In image padding, extra rows and columns are added to the image for the processing. The values in these extra rows and columns may be filled by 0s or 1s, or border rows and columns may be duplicated (Fig. 4.19). If the extra values are filled by 0s then the padding is known as zero padding. With padding, the size of an image increases. The number of rows or columns to be added depends on the size of the mask used.

In some applications, an image is required to be of a particular size. In such cases, if the image size is different, then image padding can also be used to change the size of the image.

In MATLAB, the padarray function is used for padding the image.

4.11 Histogram and Histogram Processing

A histogram is a graphical representation of the distribution of various intensity levels in an image. Mathematically, the histogram of a digital image with L intensity levels in the range [0, max gray level] is defined as the discrete function $h(r_k) = n_k$,

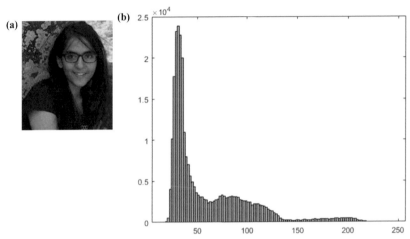

Fig. 4.20. (a) An image (b) Corresponding histogram.

where r_k is the kth intensity level in the range [0, Max gray-level] and n_k is the number of pixels in the image with intensity r_k.

A histogram provides the information regarding image contrast and brightness but does not provide any information regarding spatial relationships. In a histogram, X-axis shows the pixel intensity values and the Y-axis shows the count of these intensity values in a particular image.

A histogram can be normalized, if required, by dividing all elements of the histogram by the total number of pixels in the image.

In MATLAB, the imhist function is used to find the histogram of an image. Figure 4.20 shows an image and the corresponding histogram obtained using the imhist function. The X-axis shows the intensity levels and the Y-axis displays the number of pixels.

A histogram can be used to find the properties of an image in terms of its brightness and contrast, as shown in Fig. 4.21. In a dark image, most of the pixels are concentrated towards 0 (dark) in a histogram, as in Fig. 4.20b. In a bright image, however, most of the pixels in the histogram are found in the area towards 1. Similarly, in a low contrast image, the pixels are concentrated in a very small range of intensity values, while a high contrast image has intensity value spread over a good range of intensities. These cases are shown in Fig. 4.21.

Image enhancement in terms of brightness and contrast manipulation can be done with the help of a histogram. This process is known as histogram equalization.

4.11.1 Histogram Equalization

If the histogram of an image contains mostly dark pixels or mostly bright pixels, due to an improperly exposed photograph, for example, then the image can be

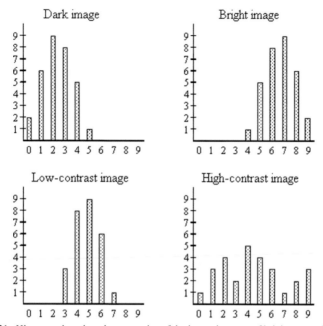

Fig. 4.21. Histogram based on the properties of the image in terms of brightness and contrast.

enhanced by histogram equalization. Histogram equalization spreads the histogram as evenly as possible over the full intensity scale. Mathematically, this is done by calculating cumulative sums of the pixel intensity values for each gray level value x in the histogram. The sum implies the number of gray levels that should be allocated to the range $[0, x]$ and is proportional to the cumulative frequency $t(x)$ and to the total number of gray levels g:

$$f(x) = g \cdot \frac{t(x)}{n} - 1 \tag{4.14}$$

Here n is the total number of pixels in the image.

In MATLAB, histogram equalization can be performed using the histeq function. Figure 4.22 shows the image 4.20a after performing histogram equalization, and with it the corresponding histogram. On comparing image 4.20a and 4.22a and the corresponding histograms 4.20b and 4.22b, the effect of histogram equalization can clearly be observed. In 4.22b, the pixel intensities are distributed along the full range of the intensity values.

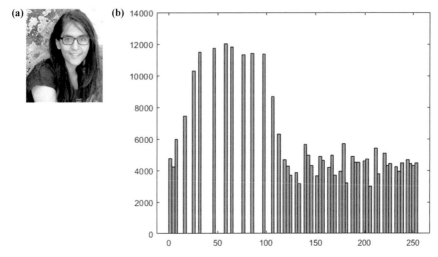

Fig. 4.22. (a) Image 4.19(a) after performing histogram equalization (b) Corresponding histogram.

Summary

- Image enhancement techniques are used to process an image in order to make it 'better' for a human viewer for a particular application.
- Image enhancement involves either of the processes: Smoothing or Sharpening.
- Spatial domain techniques operate directly on the pixel intensities in the image plane itself.
- Transform domain techniques process an image after converting it into Frequency domain.
- Point processing operations are those operations which are applied on a pixel independently of its neighbors.
- Mask processing operations consider the neighborhood pixels as well as the pixel being processed.
- Arithmetic operations: image addition, image subtraction, image multiplication and image division can be applied to digital images.
- Logical operations AND, OR and NOT can be applied on digital images. Other logical operations on the images can be applied with a combination of these basic operations.
- Image Padding is the process of adding rows and/or columns to an image for processing requirements.
- A histogram is a graphical representation of intensity distribution in an image. An image can also be processed using a histogram.

5

Image Processing in Frequency Domain

5.1 Introduction

In general, an image is represented as an array of pixels but it can also be represented as a sum of a number of sine waves of different frequencies, amplitudes and directions, known as Fourier representation or representation in Frequency domain. The parameters specifying the sine waves in Frequency domain are known as Fourier coefficients. There are some image processing activities that can be performed more efficiently by transforming an input image into a frequency domain. An image is converted into frequency domain by applying an image transform. An image transform separates components of an image according to their frequency so that the required processing can be done on these components. This process is analogous to the concept of a prism that separates a light ray into various color components. A number of image transforms are available, such as the Fourier transform, discrete cosine transform, Haar transform and Hough transform. After converting the image into frequency domain, appropriate processing is done and the inverse transformation is subsequently applied in order to get the final image (Fig. 5.1).

Frequency-domain techniques are based on the convolution theorem. Let in spatial domain, an image $f(x, y)$ is operated on by a linear operator in order to get the processed image $g(x,y)$, i.e.,

$$g(x, y) = f(x, y) * h(x, y) \tag{5.1}$$

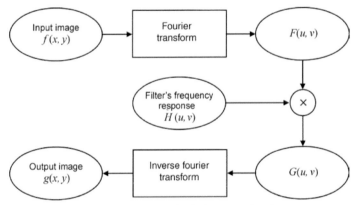

Fig. 5.1. Image processing in frequency-domain.

According to convolution theorem, corresponding to Eq. 5.1, the following relation holds in frequency-domain:

$$G(u, v) = F(u, v)\,H(u, v) \tag{5.2}$$

where G(u,v), F(u,v), and H(u,v) are the Fourier transforms corresponding to g (x,y), f(x,y), and h(x,y), respectively. This multiplication of two Fourier transforms is done point by point and therefore the dimensions of both the transforms should be equal. If the size of both transforms is not equal, then padding needs to be done before convolution. The convolution kernel H (u, v) has to be selected as per the requirement, such that the desired operation is performed.

Many image processing operations can be expressed in the form of frequency domain operations. The filter function in spatial domain can be implemented by corresponding frequency domain filters. This can be observed in Fig. 5.2, in which two frequency filters are given which correspond to spatial domain filters.

In image processing, an image is stored in spatial domain. Therefore, in order to perform processing on an image in the transform domain, initially the image has to be converted into transform domain then an algorithm is applied in the transform

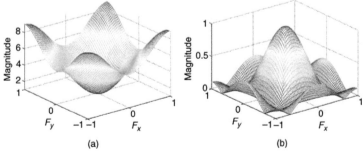

Fig. 5.2. Frequency-domain filters (a) A low-pass frequency domain filter equivalent to a 3 × 3 average filter in the spatial domain; (b) A high-pass frequency domain filter equivalent to a 3 × 3 composite Laplacian sharpening filter in the spatial domain.

domain in order to perform the required image processing, and eventually the inverse transformation is applied in order to get the image in spatial domain. Figure 5.3 shows an image and its corresponding Fourier spectrum.

The steps to be performed in frequency domain filtering are given below (Fig. 5.4):

Step 1: Apply the Fourier Transform in order to convert the image into a 2D frequency-domain.

Step 2: Apply the appropriate frequency domain filter.

Step 3: Apply the inverse Fourier Transform in order to convert the image into spatial domain.

In MATLAB®, the two-dimensional Fourier Transform of an image and its inverse can be obtained using functions fft2 and ifft2, respectively. The two-dimensional Fourier Transform is shifted in such a way that the zero-frequency component coincides with the center of the figure for visualization purposes. This

(a)   (b)

Fig. 5.3. (a) Images; (b) Fourier spectrum corresponding to the images in (a).
Original image corresponding to upper left image reprinted with permission of The MathWorks, Inc.

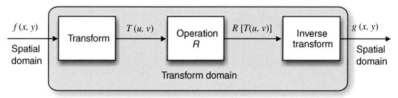

Fig. 5.4. Processing in frequency domain.

can be done using the fftshift function in MATLAB. To shift the Fourier transform back to its position, the ifftshift function is used in MATLAB.

5.2 Low-Pass Filtering in Frequency Domain

Low-pass filters attenuate the high-frequency components in the Fourier transform of an image, while low-frequency components remain unchanged. This process is the smoothing process, as described in the previous chapter. The effect of applying a low-pass filter (LPF) to an image is visible blurring. There are three common low pass filters:

  i) Ideal Low-pass filter (ILPF)
 ii) Butterworth Low-pass filter (BLPF)
iii) Gaussian Low-pass filter (GLPF)

Ideal low-pass filtering is very sharp filtering in terms of attenuating frequencies, while Gaussian low-pass filtering is very smooth. Butterworth low-pass filtering is in between these two extremes.

5.2.1 Ideal Low-pass Filter

An ideal low-pass filter (ILPF) enhances all frequency components within a specified radius (from the center of the Fourier Transform), while attenuating all others. Mathematically it is represented as:

$$H(u, v) = \begin{cases} 1 & if\, D(u, v) \leq D_0 \\ 0 & if\, D(u, v) > D_0 \end{cases} \tag{5.3}$$

where D_0 is the cut-off frequency and has a non-negative value. This cut-off frequency determines the amount of frequency components passed by the filter. If this value is small, the number of filtered image components will be larger. The value of D_0 has to be selected very carefully, as per the requirement. $D(u,v)$ is the distance of a point (u,v) from the center of the frequency rectangle.

Figure 5.5 shows a mesh and the corresponding image for an ideal low-pass filter. An Ideal low-pass filter is radially symmetric about the origin. It is difficult to implement an Ideal low-pass filter using hardware due to the sharp cut-off frequencies.

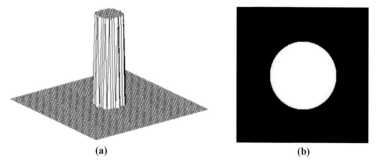

(a)   (b)

Fig. 5.5. (a) Mesh showing an ideal low pass filter (b) Filter shown in the form of an image.

The Low-pass filter transfer function shown in Fig. 5.5b can be generated using the MATLAB code below:

```
[a b]=freqspace(256, 'meshgrid');
H=zeros(256,256);
d=sqrt(a.^2+b.^2)<0.5;
H(d)=1;
figure, imshow(H)
```

Figure 5.6 shows an image and the effect of an Ideal low-pass filter in frequency domain with various radii. It can be implemented in MATLAB using the code sequence below:

```
function [out, H] = ideal_low(im, fc)
    imf = fftshift(fft2(im));
    [co,ro]=size(im);
    H = ideal_filter(co,ro,fc);
    outf=imf.*H;
    out=abs(ifft2(outf));
%Ideal low pass filter
function H = ideal_filter(co,ro,fc)
    cx = round(co/2); % finds the center of the image
    cy = round (ro/2);
    H=zeros(co,ro);
    if fc > cx & fc > cy
        H = ones(co,ro);
        return;
    end;
    for i = 1 : co
        for j = 1 : ro
```

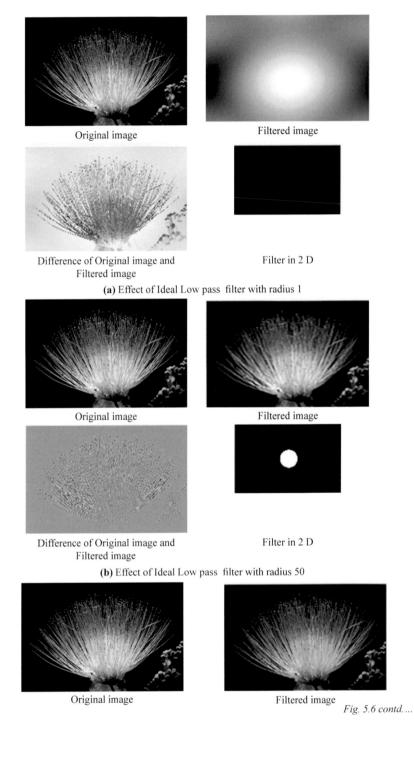

Original image

Filtered image

Difference of Original image and
Filtered image

Filter in 2 D

(a) Effect of Ideal Low pass filter with radius 1

Original image

Filtered image

Difference of Original image and
Filtered image

Filter in 2 D

(b) Effect of Ideal Low pass filter with radius 50

Original image

Filtered image

Fig. 5.6 contd....

...Fig. 5.6 contd.

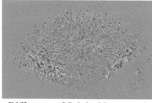 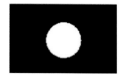

Difference of Original image and   Filter in 2D
Filtered image

(c) Effect of Ideal Low pass filter with radius 100

Fig. 5.6. An image and the effect of Ideal low-pass filters with various radii in frequency domain.

```
if (i-cx).^2 + (j-cy).^2 <= fc .^2
        H(i,j)=1;
    end;
  end;
end;
```

5.2.2 Butterworth Low-pass Filter (BLPF)

Butterworth Low-Pass Filter is a family of filters that provides a filtering technique whose behavior is a function of the cut-off frequency and the order of the filter n. Mathematically, the Butterworth filter can be represented as:

$$H(u,v) = \frac{1}{1 + [\frac{D(u,v)}{D_0}]^{2n}} \tag{5.4}$$

The shape of the response of the filter frequency can be controlled by the value of n. Higher values of n correspond to steeper transitions, approaching to ideal low-pass filter. Figure 5.7 shows a mesh showing a butter worth low pass filter and the filter in the form of an image. Figure 5.8 shows effect Butterworth filter of various radii on an image. Butterworth Low-pass filters of order 1 and order 2 do not display a ringing effect but this may be observed in higher orders. Butterworth Low-pass filters of order 2 are a good choice between effective low-pass filtering and acceptable ringing effect. It can be implemented in MATLAB using the code sequence as below:

```
% butterworth low pass filter
function [out, H] = butterworth_low (im,fc,n)
  [co,ro] = size(im);
  cx = round(co/2); % find the center of the image
  cy = round (ro/2);
```

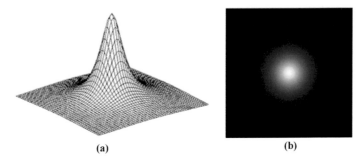

(a)   (b)

Fig. 5.7. (a) Mesh showing a Butterworth Low-pass filter (b) Filter shown in the form of an image.

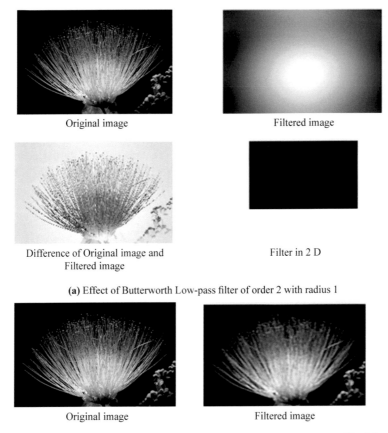

Original image   Filtered image

Difference of Original image and   Filter in 2 D
Filtered image

(a) Effect of Butterworth Low-pass filter of order 2 with radius 1

Original image   Filtered image

Fig. 5.8 contd. ...

...Fig. 5.8 contd.

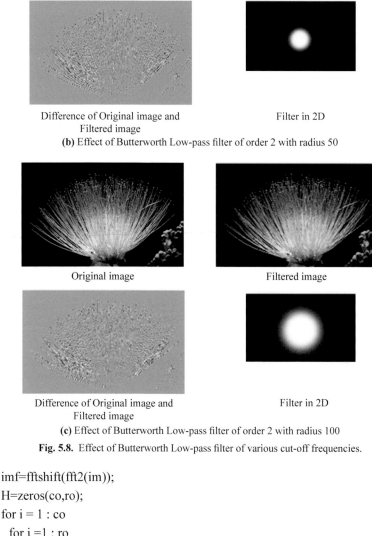

Difference of Original image and
Filtered image

Filter in 2D

(b) Effect of Butterworth Low-pass filter of order 2 with radius 50

Original image

Filtered image

Difference of Original image and
Filtered image

Filter in 2D

(c) Effect of Butterworth Low-pass filter of order 2 with radius 100

Fig. 5.8. Effect of Butterworth Low-pass filter of various cut-off frequencies.

```
imf=fftshift(fft2(im));
H=zeros(co,ro);
for i = 1 : co
  for j =1 : ro
    d = (i-cx).^2 + (j-cy).^ 2;
    H(i,j) = 1/(1+((d/fc/fc).^(2*n)));
  end;
end;
outf = imf .* H;
out = abs(ifft2(outf));
```

5.2.3 *Gaussian Low-pass Filter (GLPF)*

A Gaussian Low-pass filter uses a filter function that attenuates high frequencies and the shape of this function is based on a Gaussian curve. Gaussian filter does not have ringing effect. Mathematically, a Gaussian Low-pass filter can be represented as:

$$H(u, v) = e^{-D^2} (u, v)/2D_0^2 \tag{5.5}$$

Figure 5.9 shows a mesh of Gaussian Low-pass filter and the filter in the form of an image. Figure 5.10 shows an image and the effect of the Gaussian low pass filter on this image. It can be implemented in MATLAB using code sequence below:

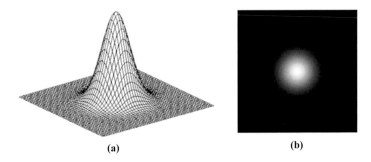

| (a) | (b) |

Fig. 5.9. (a) Mesh showing a Gaussian Low-pass filter (b) Filter shown in the form of an image.

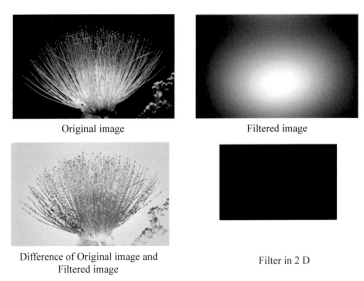

Original image   Filtered image

Difference of Original image and   Filter in 2 D
Filtered image

(a) Effect of Gaussian Low-pass filter with radius 1

Fig. 5.10 contd....

...Fig. 5.10 contd.

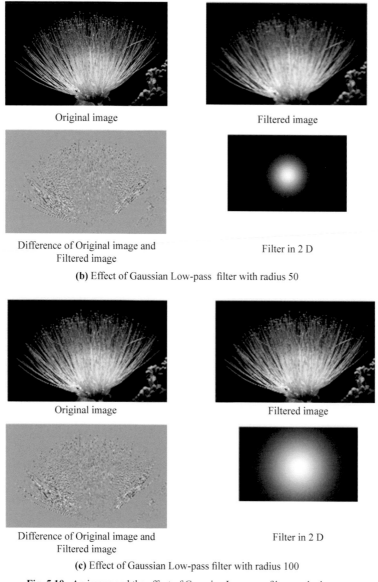

Original image

Filtered image

Difference of Original image and
Filtered image

Filter in 2 D

(b) Effect of Gaussian Low-pass filter with radius 50

Original image

Filtered image

Difference of Original image and
Filtered image

Filter in 2 D

(c) Effect of Gaussian Low-pass filter with radius 100

Fig. 5.10. An image and the effect of Gaussian Low-pass filter on the image.

```
%Gaussian lowpass filter
function [out, H] = gaussian_low(im, fc)
    imf = fftshift(fft2(im));
    [co,ro]=size(im);
```

```
    out = zeros(co,ro);
    H = gaussian_filter(co,ro, fc);
    outf= imf.*H;
    out=abs(ifft2(outf));
    % gaussian filter
function H = gaussian_filter(co,ro, fc)
    cx = round(co/2); % find the center of the image
    cy = round (ro/2);
    H = zeros(co,ro);
        for i = 1 : co
            for j = 1 : ro
                d = (i-cx).^2 + (j-cy).^2;
                H(i,j) = exp(-d/2/fc/fc);
        end;
    end;
```

5.3 High-Pass Filtering

High-pass filters, also known as sharpening filters, attenuate the low-frequency components from the Fourier transform of an image, while leaving the high-frequency components unchanged. The effect of applying a high-pass filter to an image is visible sharpening, which is opposite to smoothing. If a low-pass filter is available, then its corresponding high pass filter can be obtained by:

High-Pass filter function = 1-Low-Pass filter function   (5.6)

5.3.1 Ideal High-Pass Filter

An Ideal High-pass filter attenuates all frequency components within a specified radius (from the center of the Fourier transform), while enhancing all others. Mathematically, an Ideal High-pass Filter can be described as:

$$H(u, v) = \begin{cases} 0 & if D(u, v) \le D_0 \\ 1 & if D(u, v) > D_0 \end{cases} \qquad (5.7)$$

where D_0 is the cut-off frequency and $D(u,v)$ is the distance of a point (u,v) from the center of the frequency rectangle.

An Ideal High-Pass Filter behaves oppositely to an Ideal Low-Pass Filter as it sets all frequencies inside the defined radius to zero, while passing frequencies outside the circle. It is not possible to implement IHPF in hardware. Figure 5.11 shows a mesh corresponding to an Ideal High-pass filter and the corresponding filter in image form. Figure 5.12 shows the effects of Ideal High-pass filters of

(a)

(b)

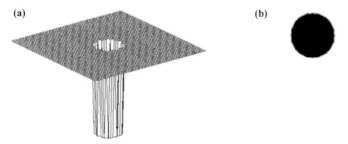

Fig. 5.11. (a) Mesh showing an Ideal High-pass filter (b) Filter shown in the form of an image.

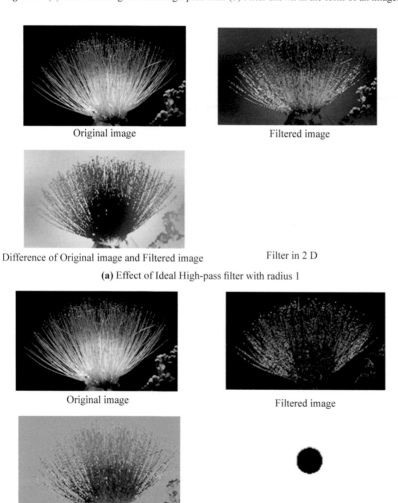

Original image

Filtered image

Difference of Original image and Filtered image

Filter in 2 D

(a) Effect of Ideal High-pass filter with radius 1

Original image

Filtered image

Difference of Original image and Filtered image

Filter in 2 D

(b) Effect of Gaussian High pass filter with radius 50

Fig. 5.12 contd....

...Fig. 5.12 contd.

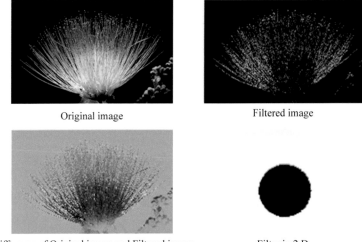

Original image

Filtered image

Difference of Original image and Filtered image

Filter in 2 D

(c) Effect of Ideal High-pass filter with radius 100

Fig. 5.12. An image and the effect of Ideal High-pass filters on the image.

various radii on an image. It can be implemented in MATLAB using the code sequence below:

```
% Ideal High pass filter
function [out, H] = ideal_high(im, fc)
    imf = fftshift(fft2(im));
    [co,ro]=size(im);
    H = ideal_filter(co,ro,fc);
    H = 1-H;
    outf=imf.*H;
    out=abs(ifft2(outf));
```

5.3.2 Butterworth High-Pass Filter

The Butterworth High-pass filter can be mathematically described as

$$H(u,v) = \frac{1}{1 + [\frac{D_0}{D(u,v)}]^{2n}} \qquad (5.9)$$

where D_0 is the cut-off frequency and n is the order of the filter.

Butterworth High-pass filters are smoother in comparison to Ideal High-pass filters. Butterworth High-pass filters of order 2 are a good choice between Ideal and Gaussian High-pass filters. Figure 5.13 shows a mesh of Butterworth High-pass filter and the corresponding image. Figure 5.14 shows the effects of Butterworth High-pass filters of various radii and of order 2 on an image. It can be implemented in MATLAB using the code sequence below:

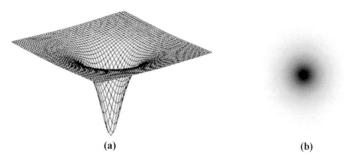

(a)   (b)

Fig. 5.13. (a) Mesh showing Butterworth High-pass filter (b) Filter shown in the form of an image.

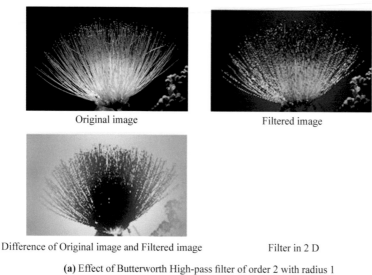

Original image   Filtered image

Difference of Original image and Filtered image   Filter in 2 D

(a) Effect of Butterworth High-pass filter of order 2 with radius 1

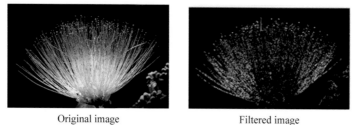

Original image   Filtered image

Fig. 5.14 contd. ...

...Fig. 5.14 contd.

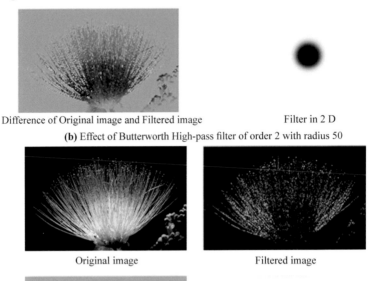

Difference of Original image and Filtered image   Filter in 2 D

(b) Effect of Butterworth High-pass filter of order 2 with radius 50

Original image   Filtered image

Difference of Original image and Filtered image   Filter in 2 D

(c) Effect of Butterworth High pass filter of order 2 with radius 100

Fig. 5.14. An image and the effect of Butterworth High-pass filters of order 2 on the image.

```
% butterworth high pass filter
function [out, H] = butterworth_high (im,fc,n)
    [co,ro] = size(im);
    cx = round(co/2); % find the center of the image
    cy = round (ro/2);
    imf=fftshift(fft2(im));
    H=zeros(co,ro);
    for i = 1 : co
        for j =1 : ro
            d = (i-cx).^2 + (j-cy).^ 2;
            if d ~= 0
            H(i,j) = 1/(1+((fc*fc/d).^(2*n)));
        end;
```

```
    end;
end;
outf = imf .* H;
out = abs(ifft2(outf));
```

5.3.3 Gaussian High-Pass Filter

A Gaussian High-pass filter attenuates low frequencies from the Fourier transform by using a filter function whose shape is based on a Gaussian curve. The Gaussian High-pass Filter can be mathematically represented as:

$$H(u, v) = 1 - e^{-D^2(u,v)/2D_0^2} \qquad (5.8)$$

Figure 5.15 shows a mesh corresponding to a Gaussian High-pass filter and the filter in image form. Figure 5.16 shows the effect of Gaussian High-pass filters of various radii on an image. It can be implemented using a MATLAB code similar to:

```
%Gaussian high pass filter
function [out, H] = gaussian_high(im, fc)
    imf = fftshift(fft2(im));
    [co,ro]=size(im);
    H = gaussian_filter(co,ro, fc);
    H = 1-H;
    out = zeros(co,ro);
    outf= imf.*H;
    out=abs(ifft2(outf));
```

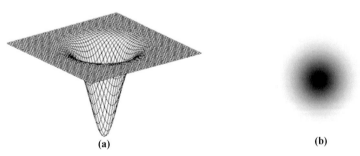

(a)   (b)

Fig. 5.15. (a) Mesh showing Gaussian High-pass filter (b) Filter shown in the form of an image.

5.4 High-Frequency Emphasis Filter

High-frequency emphasis filtering is a technique that preserves the low-frequency contents of the input image and, at the same time, enhances its high-frequency components. In a high frequency emphasis filter, the high-pass filter function is multiplied by a constant and an offset is added to the result, i.e.,

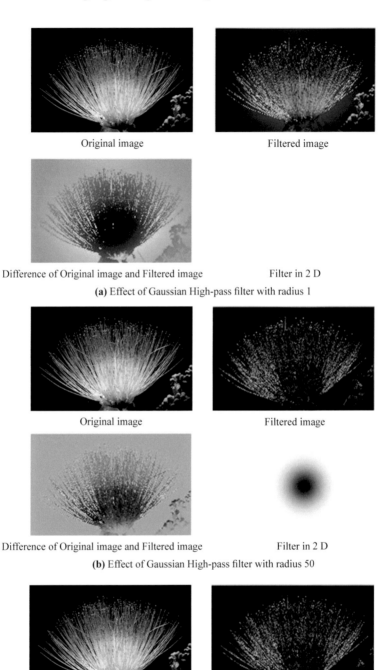

Original image

Filtered image

Difference of Original image and Filtered image

Filter in 2 D

(a) Effect of Gaussian High-pass filter with radius 1

Original image

Filtered image

Difference of Original image and Filtered image

Filter in 2 D

(b) Effect of Gaussian High-pass filter with radius 50

Original image

Filtered image

Fig. 5.16 contd....

...Fig. 5.16 contd.

Difference of Original image and Filtered image   Filter in 2 D

(c) Effect of Gaussian High-pass filter with radius 100

Fig. 5.16. An image and the effect of Gaussian High-pass filters on the image.

High Frequency Emphasis Filter = a + b High-Pass Filter   (5.10)

where a is the offset and b is the multiplier. In this equation, the multiplier b enhances the high-frequency components.

Summary

- An image can be represented in Frequency domain as a sum of a number of sine waves of different frequencies, amplitudes and directions known as Fourier representation.
- Low-pass filters attenuate the high-frequency components in the Fourier transform of an image, without affecting low-frequency components.
- High-pass filters attenuate the low-frequency components in the Fourier transform of an image, without affecting high-frequency components.
- High-frequency emphasis filtering preserves the low-frequency contents of the input image and, at the same time, enhances its high-frequency components.

6

Image Denoising

6.1 Introduction

These days, images and videos are part of our daily life. There are many applications which require images to be visually impressive and clear. There is a possibility that a digital image may get corrupted by noise during transmission or during the process of image acquisition by cameras. This noise degrades the quality of an image in terms of visual appearance.

The process of removing noise from a digital image is known as image denoising. In image denoising, an original image has to be estimated from a noisy image by removing noise. This is a complex process. In this process, the maximum amount of noise must be removed without eliminating the most representative characteristics of the image, such as edges, corners and other sharp structures. A number of image denoising techniques are available that can be used to reduce the effect of noise caused by any noise sources.

As shown in Fig. 6.1, an input image $f(x, y)$ becomes a noisy image $g(x, y)$ with the addition of noise $n(x, y)$. Given noisy image $g(x, y)$ and some information about the noise $n(x, y)$, the objective of the denoising process is to obtain an approximate $\hat{f}(x, y)$ of the original image. This approximate image should be as similar to the original image as possible and, in general, the more information about $n(x, y)$ is available, the closer $\hat{f}(x, y)$ will be to $f(x, y)$. In the next section, a brief review of some common noise types and some conventional noise reduction filters are described.

6.2 Image Noise Types

Noise in an image is caused by random variations of brightness or color information which are not part of original image. Noise in digital images is introduced during the process of image acquisition and/or transmission from one location to other.

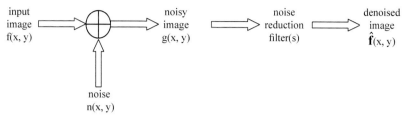

Fig. 6.1. A model of the image denoising process.

There are two categories of image noise: spatially independent noise and spatially dependent noise.

6.2.1 *Spatially Independent Noise*

Spatially independent noise is independent of spatial coordinates and uncorrelated with pixel values. Some commonly occurring spatially independent noise types, with their respective mean and variances, are given in Table 6.1.

Table 6.1. Some common image noise types.

Name	PDF	Mean and Variance
Gaussian Noise	$p(z) = \dfrac{1}{\sqrt{2\pi}b} e^{\frac{-(z-a)^2}{2b^2}}, \qquad -\infty < z < \infty$	$m = a, \sigma^2 = b^2$
Salt and pepper (impulse) Noise	$p(z) = \begin{cases} P_a, & z = a \\ P_b, & z = b \\ 0, & \text{otherwise} \end{cases}$	$m = aP_a + bP_b,$ $\sigma^2 = (a-m)^2 P_a + (b-m)^2 P_b$
Rayleigh Noise	$p(z) = \begin{cases} \dfrac{2}{b}(z-a)e^{\frac{-(z-a)^2}{b}}, & z \geq a \\ 0, & z < a \end{cases}$	$m = a + \sqrt{\dfrac{\pi b}{4}}, \sigma^2 = \dfrac{b(4-\pi)}{4}$
Gamma Noise	$p(z) = \begin{cases} \dfrac{a^b z^{b-1}}{(b-1)!}e^{-az}, & z \geq 0 \\ 0, & z < 0 \end{cases}$	$m = \dfrac{b}{a}, \sigma^2 = \dfrac{b}{a^2}$
Exponential Noise	$p(z) = \begin{cases} ae^{-az}, & z \geq 0 \\ 0, & z < 0 \end{cases}$	$m = \dfrac{1}{a}, \sigma^2 = \dfrac{1}{a^2}$
Uniform Noise	$p(z) = \begin{cases} \dfrac{1}{b-a}, & \text{if } a \leq z \leq b \\ 0, & \text{otherwise} \end{cases}$	$m = \dfrac{a+b}{2}, \sigma^2 = \dfrac{(b-a)^2}{12}$

6.2.2 Spatially Dependent Noise

Spatially dependent noise depends on spatial coordinates and has some correlation with pixel values. A common spatially dependent noise is periodic noise which is generally caused by electrical or electromechanical interferences during image acquisition.

In MATLAB®, the command to add noise to an intensity image is: imnoise(I,*type*), where *type* specifies the type of noise:

'gaussian' - Gaussian white noise with constant mean and variance.

'localvar'- zero-mean Gaussian white noise with an intensity-dependent variance.

'poisson' - Poisson noise.

'salt & pepper' - on and off pixels.

'speckle' - multiplicative noise.

6.3 Image Denoising

The process of removing noise from an image is known as Image denoising. Based on the domain, image denoising methods are classified as spatial domain methods and transform domain methods.

6.3.1 Spatial Domain Methods

Spatial filters are generally used to remove noise from an image. A spatial filter is applied on the noisy image in order to denoise it. These filters are direct and fast processing. Spatial filters are of two types: linear spatial domain filters and non-linear spatial domain filters.

6.3.2 Transform Domain Methods

Transform domain denoising filtering methods are applied on the image transforms. Initially, a noisy image is converted into transform domain and then denoising techniques are applied on the transform of the noisy image. Depending upon the selection of basis transform functions, transform domain denoising techniques are divided into two categories: data adaptive transform or non-data adaptive transform domain denoising techniques.

- **Data Adaptive Transform**
- **Non-Data Adaptive Transform**

Figure 6.2 shows an image and the corresponding image after adding Gaussian noise and Salt and Pepper noise of various densities. The MATLAB commands to obtain these noisy images are also given in below figures.

Reducing noise from an image is a standard problem in digital image processing. In an image, the edges of objects are very important for a clear visual

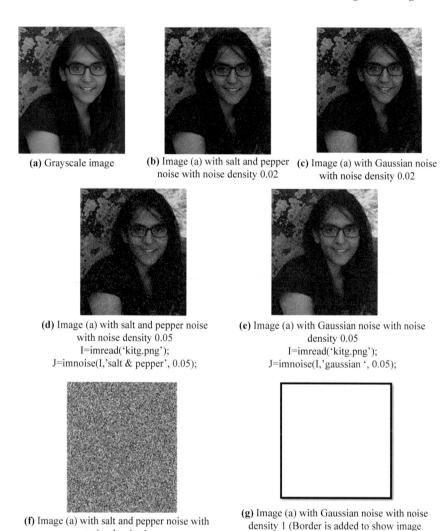

(a) Grayscale image

(b) Image (a) with salt and pepper noise with noise density 0.02

(c) Image (a) with Gaussian noise with noise density 0.02

(d) Image (a) with salt and pepper noise with noise density 0.05
I=imread('kitg.png');
J=imnoise(I,'salt & pepper', 0.05);

(e) Image (a) with Gaussian noise with noise density 0.05
I=imread('kitg.png');
J=imnoise(I,'gaussian ', 0.05);

(f) Image (a) with salt and pepper noise with noise density 1
I=imread('kitg.png');
J=imnoise(I,'salt & pepper', 1);

(g) Image (a) with Gaussian noise with noise density 1 (Border is added to show image dimensions)
I=imread('kitg.png');
J=imnoise(I,'gaussian ', 1);

Fig. 6.2. An image and the image after introduction of 'salt and pepper' and 'Gaussian' noise of various densities.

appearance. The denoising process is basically a smoothing process, so, as a side effect of the denoising process, edges, corners and sharp structures in the images are also smoothed or blurred. Therefore, it is very important in denoising techniques that the noise should be removed while preserving the edges, corners and other sharp structures in the image.

A simple denoising technique, known as Linear Translation Invariant (LTI) filtering, is implemented using a convolution mask. For example, box filtering, also known as mean filtering or image averaging, is implemented by a local averaging operation where the value of each pixel is replaced by the average of all the values in the local neighborhood. These techniques have been discussed with regard to the smoothing of an image in earlier chapters of the book. Box filter techniques are very quick, but their smoothing effect is often unsatisfactory. Some LTI filtering techniques used for denoising do not calculate the mean of a neighborhood, e.g., Gaussian smoothing denoising and Weiner filter technique.

LTI denoising techniques remove the noise but also blur sharp structures in the image, which is undesirable. To reduce the undesirable effects of linear filtering, a number of edge-preserving denoising techniques have been proposed.

A number of non-linear denoising techniques are available, e.g., median, weighted median, rank conditioned rank selection, which perform better than the Linear denoising techniques.

A number of techniques which are non-linear in nature and consider local geometries in the denoising process are researched. Generally, these techniques are based on Partial Differential Equations (PDEs) and variation models.

A simple, non-iterative, local filtering method known as the bilateral filter is simple, but its direct implementation is slow. The brute force implementation is of $O(Nr^2)$ time, which is prohibitively high if the kernel radius r is large.

A guided filter can resolve the issues raised by a bilateral filter and can perform effective edge-preserving denoising by considering the content of a guidance image. Unlike the bilateral filter, the guided filter avoids the gradient reversal artefacts that could appear in detail enhancement and HDR compression.

Techniques based on Singular Value Decomposition (SVD) are also used in image noise filtering. Apart from the edge-preserving filters mentioned so far, wavelets also gave superior performance in edge-preserving denoising due to properties such as sparsity and multiresolution structure.

In MATLAB, various simple filters are available that can be used to remove noise. For complex denoising techniques, programs can be developed in MATLAB, but some filters like the Averaging filter, Median filters, an adaptive linear filter (i.e., Weiner filter) can be used simply using MATLAB functions fspecial and imfilter. A MATLAB code similar to the code below can be used in order to denoise an image by applying filters.

```
Img = imread(image name);
%Image with noise
a = imnoise(Img,'Gaussian',0.02);
%Average filter of size 3x3
H = fspecial('Average',[3 3]);
b = imfilter(a,H);
%Average filter of size 3x3
```

H = fspecial('Average',[5 5]);
c = imfilter(a,H);
%Average filter of size 3x3
H = fspecial('Average',[7 7]);
d = imfilter(a,H);
%Average filter of size 3x3
H=fspecial('Average',[9 9]);
e=imfilter(a,H);
subplot(3,2,1),imshow(Img),title('Original image')
subplot(3,2,2),imshow(a),title('Gaussian noise with mean 0.02')
subplot(3,2,3),imshow(uint8(b)),title('3 x 3 Average filter')
subplot(3,2,4),imshow(uint8(c)),title('5 x 5 Average filter')
subplot(3,2,5),imshow(uint8(d)),title('7x7 Average filter')
subplot(3,2,6),imshow(uint8(d)),title('9x9 Average filter')

Figures 6.3–6.6 show an image, its noisy version with Gaussian and Salt and Pepper noise with various mean and effect of applying various filters.

Original image

Gaussian noise with mean 0.02

3 x 3 Average filter

5 x 5 Average filter

7x7 Average filter

9x9 Average filter

Fig. 6.3. An image, its noisy version with Gaussian noise having mean 0.02 and effect of applying Average filters of various size.

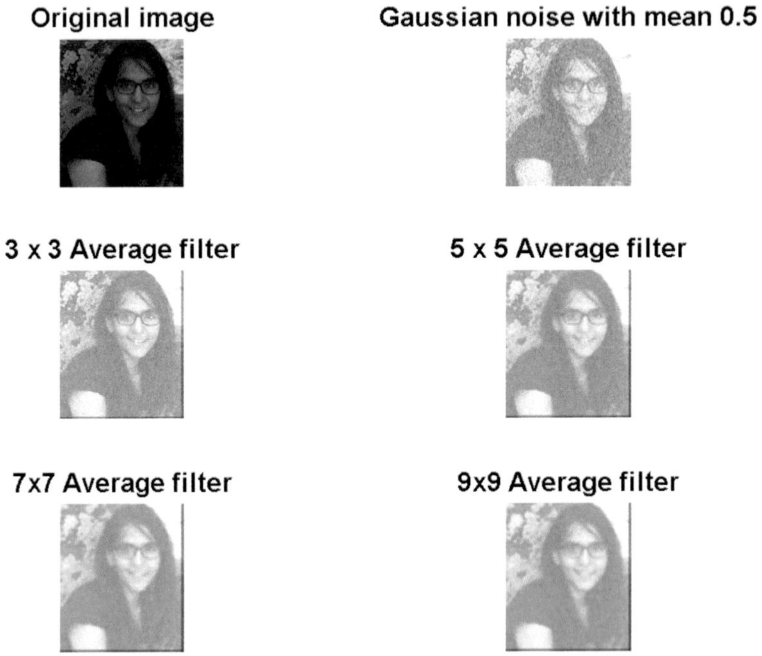

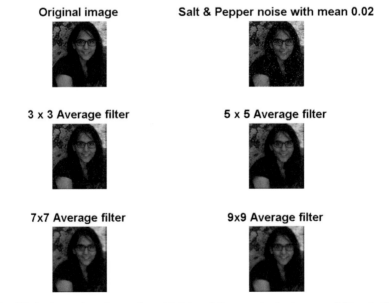

Original image

Gaussian noise with mean 0.5

3 x 3 Average filter

5 x 5 Average filter

7x7 Average filter

9x9 Average filter

Fig. 6.4. An image, its noisy version with Gaussian noise having mean 0.5 and effect of applying Average filters of various size.

Original image

Salt & Pepper noise with mean 0.02

3 x 3 Average filter

5 x 5 Average filter

7x7 Average filter

9x9 Average filter

Fig. 6.5. An image, its noisy version with Salt and Pepper noise having mean 0.02 and effect of applying Average filters of various size.

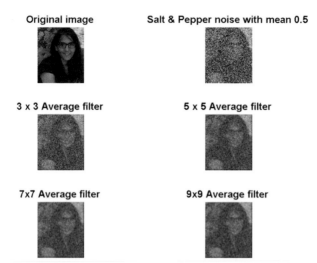

Fig. 6.6. An image, its noisy version with Salt and Pepper noise having mean 0.05 and effect of applying Average filters of various size.

6.5 Performance Evaluation of Denoising Techniques

It is important to evaluate the performance of a denoising technique. Various parameters are available for this purpose. Let g be a noisy image corresponding to an unknown noise-free image f. By applying an image denoising algorithm, the image $\hat{f}$ is obtained. This image should be similar to f.

To evaluate the performance of a denoising technique, a parameter called Mean Square Error (MSE) is used, which is:

$$\text{MSE} = \frac{1}{M \times N} \left\| f - \hat{f} \right\|^2 = \frac{1}{M \times N} \sum_{x=1}^{M} \sum_{y=1}^{N} \left[f(x,y) - \hat{f}(x,y) \right]^2 \qquad (6.1)$$

where $\|\cdot\|^2$ is the Euclidean norm.

Another evaluation parameter is Peak Signal-to-Noise ratio (PSNR) for 8-bit grayscale images:

$$\text{PSNR} = 10 \log_{10} \left(\frac{255^2}{\text{MSE}} \right) \qquad (6.2)$$

The value of PSNR, which is defined in decibels (dB), is used to evaluate the performance of a denoising algorithm. If the value of PSNR is high, then the performance of denoising techniques is also high; otherwise, the performance is low.

These two parameters, MSE and PSNR are easy to calculate and have clear physical meanings which are also mathematically convenient in the context of optimization. However, these are not very well matched for perceiving visual quality, i.e., these metrics do not directly measure the visual features of the resulting image. Images with identical MSE or PSNR values may be very different in their visual quality.

A parameter used to compare two images is the Structural Similarity Index (SSIM), which correlates more appropriately with the human perception. This parameter maps two images into an index in the interval $[-1, 1]$, where higher values are assigned to more similar pairs of images X and Y, calculated as

$$\text{SSIM}(X,Y) = \frac{(2\mu_X\mu_Y + C_1)(2\sigma_{XY} + C_2)}{(\mu_X^2 + \mu_Y^2 + C_1)(\sigma_X^2 + \sigma_Y^2 + C_2)} \tag{6.3}$$

where μ_x, μ_y, σ_X^2 and σ_Y^2 are the averages and variances of X and Y, σ_{XY} is the covariance between X and Y, and both C_1 and C_2 are predefined constants.

In order to measure the performance of an edge-preserving denoising filter, another parameter, known as Pratt's figure of merit (FOM), is also used. The value of this metric indicates how much the detected edges in the filtered image coincide with the edges in original image. In other words, the value of FOM gives us an idea about how effectively edges of an original image are preserved in the filtered image. Figure of Merit (FOM) is defined as:

$$\text{FOM} = \frac{1}{\max(N_{Ide}, N_{Det})} \sum_{i=1}^{N_{Det}} \frac{1}{1 + \alpha d_i^2} \tag{6.4}$$

where, N_{Ide} and N_{Det} are the numbers of ideal and detected edge pixels, respectively, d_i is the distance between an edge point and the nearest ideal edge pixel and α is an empirical calibration constant (often 1/9) used to penalize displaced edges. The value of FOM is a number in the interval $[0, 1]$, where 1 represents the optimal value, i.e., the detected edges coincide with the ideal edges.

Summary

- A digital image may get corrupted by noise during transmission or during the process of image acquisition by cameras. This noise degrades the quality of an image in terms of visual appearance.
- Noise in an image is caused by random variations of brightness or color information which is not part of original image.
- The process of removing noise from a digital image is known as image denoising.
- Image noise can be either spatially independent noise or spatially dependent noise.
- Common image noise types are Salt and Pepper, Gaussian, Uniform, Rayleigh and Exponential noise.
- The denoising process is basically a smoothing process.
- The effectiveness of a denoising method is evaluated using MSE, PSNR, FOM, etc.

7

Image Segmentation

7.1 Introduction

Image segmentation is the process that divides an image into different regions, such that each region is homogeneous according to some well-defined features or characteristics. This is an essential and critical step of an image analysis system. It is one of the most important and difficult tasks in image processing, and the accuracy of the final result of image analysis depends on the accuracy of the image segmentation. Some common examples of image segmentation are: in image analysis for automated inspection of electronic assemblies, to divide terrain into forest areas and water bodies in satellite images and analysis of medical images obtained from MRI and X-ray machines for the purpose of identifying organs, bones, etc.

Image segmentation is one of the most essential tools in numerous image processing and computer vision tasks. Various researchers have defined image segmentation in their own way. Some formal definitions of image segmentation are as follows:

- "In computer vision, image segmentation is the process of partitioning a digital image into multiple segments (sets of pixels, also known as super-pixels)."
- "Segmentation is a process of grouping together pixels that have similar attributes."
- "Image Segmentation is the process of partitioning an image into non-intersecting regions such that each region is homogeneous and the union of no two adjacent regions is homogeneous."

In general, image processing activities can be categorized in three levels, according to types of input and output; low-level image processing, mid-level image processing and high-level image processing. Image segmentation is an important part of the last two categories. Hence, segmentation plays a crucial role in image analysis and computer vision.

7.2 Techniques of Image Segmentation

Image segmentation has no single standard technique and it is very difficult to achieve in non-trivial images. A large number of image segmentation techniques have been suggested by researchers. Segmentation techniques are based on two properties of intensity values of regions: discontinuity and similarity. In discontinuity-based techniques, an image is segmented based on the abrupt changes in the intensity, while similarity-based techniques partition the image into various regions based on similarity according to a set of predefined criteria. Figure 7.1 shows classification of segmentation techniques.

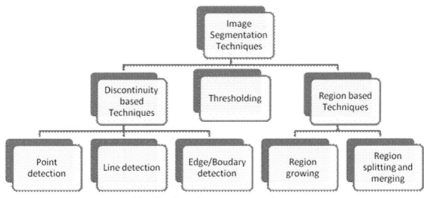

Fig. 7.1. Classification of image segmentation techniques.

7.3 Discontinuity-based Image Segmentation Techniques

In this category of image segmentation techniques, an image is divided into segments based on the abrupt changes in the intensity. A number of techniques are available for detecting basic gray level discontinuities; points, lines and edges in a digital image.

7.3.1 Point Detection

An isolated point can be defined as a point which has an intensity level significantly different from its neighbors and is located in a homogeneous or nearly homogeneous area. An isolated point can be detected using masks, as shown in Fig. 7.2. A point is detected using mask processing if, at the location on which the mask is centered, the processed value is greater than some non-negative threshold. Basically, in this process, weighted differences between the center point and its neighbors are

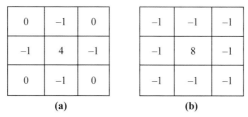

Fig. 7.2. 3×3 Laplacian masks to detect points in an image.

measured. An isolated point will be quite different from its surroundings in terms of its intensity value, and, thus, can easily be singled out using this type of mask.

So, in order to segment points in an image, the process is:

i) First apply a mask similar to Fig. 7.2
ii) Perform thresholding

In MATLAB®, the imfilter function is used for applying the mask. This can be done in MATLAB using code similar to:

```
m1_point_lapl=[−1 −1 −1;−1 8 −1; −1 −1 −1];
I = imread('input_image.jpg');
figure, imshow(I)
title('Original Image')

I1 = imfilter(I,m1_point_lapl);
figure, imshow(I1)
title('Result Image')
I2= im2bw(I2,0.9)
figure, imshow(I2)
title('Image after thresholding')
```

An example of point detection using this code is shown in Fig. 7.3.

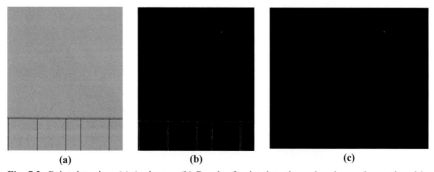

Fig. 7.3. Point detection: (a) An image, (b) Result of point detection using the mask mentioned in Fig. 7.2b, (c) Result after applying thresholding.

The value of threshold value should be selected very carefully so that the required points can be detected with various intensities.

Another technique that can be used for point detection is to find all the points in the neighborhood of size mxn for which the difference of the maximum and minimum pixel intensity exceeds a specified threshold value.

7.3.2 Line Detection

In many image processing applications, there is a need to detect lines. In such applications, the masks given in Fig. 7.4 can be used. These masks can be used to detect horizontal, vertical, 45 degree and 135 degree lines.

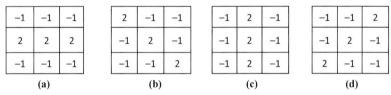

Fig. 7.4. Masks to detect single pixel wide line; (a) mask to detect horizontal lines, (b) mask to detect lines oriented with 135°, (c) mask to detect vertical lines, (d) mask to detect lines oriented with 45°.

In general, to segment lines with various orientations in an image, every mask is operated over the image and then the responses are combined:

$$R(x, y) = \max(|R1\ (x, y)|, |R2\ (x, y)|, |R3\ (x, y)|, |R4\ (x, y)|) \tag{7.1}$$

If $R(x, y) > T$, then discontinuity

An example of line detection is given in Fig. 7.5. Figure 7.5a shows an image which has lines with various orientations. This image is processed with various masks, and the effect is shown in Figs. 7.5b to e.

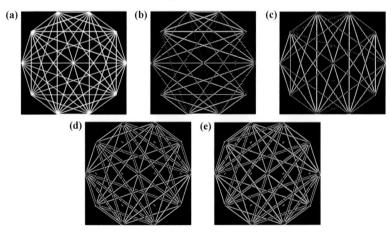

Fig. 7.5. Line detection: (a) An image, (b) Horizontal lines highlighted, (c) Vertical lines highlighted (d) lines oriented at 45° lines highlighted (e) lines oriented at 135° lines highlighted.

In MATLAB, line detection can be performed using code similar to:

```
m2_horiz=[−1 −1 −1;2 2 2; −1 −1 −1];
m2_45=[−1 −1 2;−1 2 −1; 2 −1 −1];
m2_vert=[−1 2 −1;−1 2 −1; −1 2 −1];
m2_135=[2 −1 −1;−1 2 −1; −1 −1 2];
I = imread('input_image.jpg');
figure, imshow(I)
title('Original Image')
I2 = imfilter(I, m2_horiz);
I3 = imfilter(I, m2_vert);
I4 = imfilter(I, m2_45);
I5 = imfilter(I, m2_135);
figure, imshow(I2)
title('Horizontal lines highlighted')
figure, imshow(I3)
title('Vertical lines highlighted')
figure, imshow(I4)
title('45° lines highlighted')
figure, imshow(I5)
title('135° lines highlighted')
```

Figure 7.6 shows the effect of the mask on lines in the image mathematically and Fig. 7.7 shows the complete process for finding lines with various orientations mathematically.

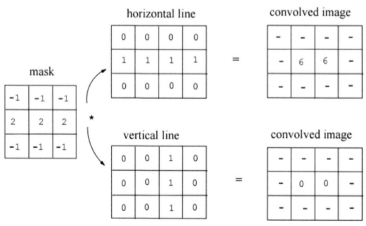

Fig. 7.6. Effect of mask line detection.

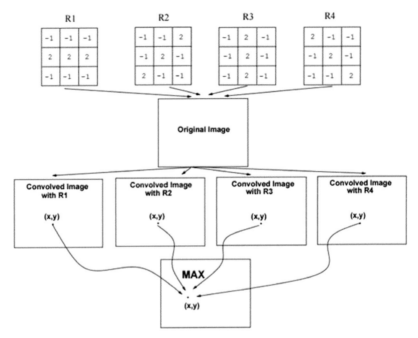

Fig. 7.7. Process of line detection in an image.

7.3.3 Edge Detection

An edge is defined as a sudden change in intensity, i.e., edges have different pixel intensity values in comparison to their neighbors. An edge is defined as the change in the intensity of an image from low to high, or vice versa. The edge is regarded as the boundary between two dissimilar regions. The edges of various objects in an image are very important as edges hold a good amount of information in an image. The edges indicate the location of objects in the image and their shape and size.

There are several applications for detection of edges of objects in an image. For example, digital artists use edge detection techniques to create dazzling image outlines. The output of an edge detection technique can be added back to an original image in order to enhance the edges.

There are an infinite number of edge orientations, widths and shapes. Edges may be straight or curved, with varying radii. There are a number of edge detection techniques to go with all these edges, each having their own strengths. Some edge detectors may work well for one application, but poorly for other applications. The choice of an edge detection technique completely depends on the application.

In general, edges are of four types; step, line, ramp and roof edge. Step edges are those edges where the image intensity changes abruptly from a value on one side of the discontinuity to a different value on the opposite side. Due to smoothing, step edges become ramp edges and line edges become roof edges, hence, sharp changes become somewhat smooth. These edge shapes are shown in Fig. 7.8.

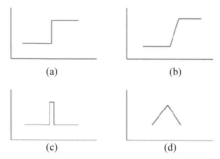

Fig. 7.8. Types of edges: (a) Step edge (b) Ramp edge (c) Line edge (d) Roof edge.

A number of edge detection techniques are available, which are described next.

7.3.3.1 Gradient-based edge detection

The first order derivative in image processing is implemented using the magnitude of the gradient. If the image is represented by the function f(x,y), then the gradient of f at coordinates (x,y) can be computed as:

$$\nabla f = grad(f) = \begin{bmatrix} G_x \\ G_y \end{bmatrix} = \begin{bmatrix} \dfrac{\partial f}{\partial x} \\ \dfrac{\partial f}{\partial y} \end{bmatrix} \tag{7.2}$$

where G_x and G_y are the gradient components in x and y directions respectively.

This gradient vector, denoted by ∇f, is a very appropriate tool for representing the strength and direction of the edge. Formally, ∇f can be defined as the geometrical property that points in the direction of the greatest rate of change of f at coordinates (x,y). The magnitude (length) and direction (angle) of ∇f, can be calculated as:

$$|\nabla f| = mag(\nabla f) = mag(x,y) = \sqrt{G_x^2 + G_y^2} \tag{7.3}$$

$$\theta(x, y) = \tan^{-1}\left(\frac{|G_y|}{|G_x|}\right) \tag{7.4}$$

$|\nabla f|$, denotes the value of the rate of change in the direction $\theta(x,y)$ of the gradient vector. It should be noted that G_x, G_y, $\theta(x, y)$, and $mag(\nabla f)$ are the images, whose sizes are identical to the original image f(x,y). In general, image $mag(\nabla f)$ is termed as 'gradient image' or simply gradient. A gradient-based edge detector method includes Roberts, Prewitt and Sobel operators.

- **Roberts edge detection (Roberts Cross Edge Detector):** The Roberts operator is a very simple and quick way to compute the approximation of the gradient of an image. The Roberts operator consists of a pair of 2 × 2 masks (Fig. 7.9) in the x and y directions. These masks are designed in such a way that preference is given to the diagonal edges. The x component (denoted by G_x) of the Robert operator detects edges oriented at 45°, while the y component (denoted by G_y)

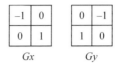

Gx   Gy

Fig. 7.9. Roberts mask.

highlights edges oriented at 135°. The result of the Roberts edge detection is shown in Fig. 7.12b.

In the Roberts edge detection technique, only four pixels' values are processed in order to get the value at an output pixel and the mathematical operations are limited to addition and subtraction. Therefore, this is very quick technique. The disadvantage of this technique is its noise sensitivity which is limited due to the small kernel. If the edges are not very sharp, the output from the Roberts edge detection is unsatisfactory.

- **The Prewitt edge detector:** The Prewitt operator is also a gradient-based edge detector, it consists of two 3×3 masks (Fig. 7.10). Basically, the Prewitt operator detects horizontal and vertical edges. The x component (denoted by G_x) of the Prewitt operator detects horizontal edges, while the y component (denoted by G_y) detects vertical edges. The results of the Prewitt edge detection are shown in Fig. 7.12c. It is computationally less expensive and a quicker method for edge detection but this technique does not work well with noisy and low contrast images.

- **The Sobel edge detector:** The Sobel operator is also based on gradient, and very similar to the Prewitt operator. The difference between the two is that the Sobel operator has '2' and '–2' values in the center of the first and third columns of the horizontal mask and first and third rows of the vertical mask (Fig. 7.11).

-1	-1	-1
0	0	0
1	1	1

Gx

-1	0	1
-1	0	1
-1	0	1

Gy

Fig. 7.10. Prewitt mask.

-1	-2	-1
0	0	0
1	2	1

Gx

-1	0	1
-2	0	2
-1	0	1

Gy

Fig. 7.11. Sobel mask.

Due to this change, more weightage is assigned to the pixel values around the edge region, and therefore, the edge intensity is increased. Similar to Prewitt,

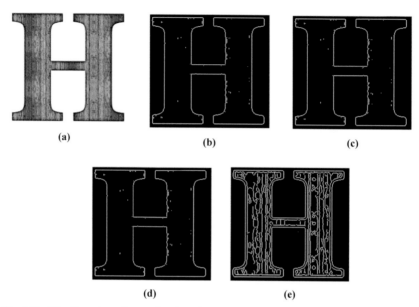

(a)

(b)

(c)

(d)

(e)

Fig. 7.12. Edge Detection using gradient-based techniques: (a) An image, (b) Result of Roberts mask, (c) Result of Prewitt mask, (d) Result of Sobel mask, (e) Result of Canny edge detector.

the Sobel operator also highlights horizontally and vertically oriented edges. The results of the Sobel edge detection are shown in Fig. 7.12d.

Due to the larger masks, the Sobel operator smooths the input image to a greater extent and is less sensitive to noise. The operator also generally produces considerably higher output values for similar edges, when compared with the Roberts Cross.

The Sobel operator is slower than the Roberts Cross operator. Due to the smoothing effect of the Sobel operator, some natural edges are converted to thicker edges. Some thinning may, therefore, be required in order to counter this.

The Prewitt edge detection technique overcomes the problem faced in the Sobel edge detection technique due to the absence of the smoothing modules. The operator adds a vector value in order to provide smoothing.

The MATLAB code for edge detection using gradient-based filters is similar to:

```
I = imread('input_image.jpg');
figure, imshow(I) % Input image
I1 = edge(I, 'Priwitt');
I2 = edge(I, 'Sobel');
```

I3 = edge(I, 'Roberts');
figure, imshow(I1) % Prewitts mask
figure, imshow(I2) % Sobel mask
figure, imshow(I3) % Roberts mask

Canny edge detection algorithm: The Canny edge detection algorithm was proposed by John F. Canny in 1986. The Canny edge detection algorithm is a widely used method in research. This edge detection method is optimal for the following three criteria:

1. Detection: The probability of detecting real edge points should be maximized while the probability of falsely detecting non-edge points should be minimized. This entails maximizing the signal-to-noise ratio of the gradient.
2. An edge localization factor: The detected edges should be very close to the real edges.
3. Number of responses: One real edge should not result in more than one detected edge, i.e., minimizing multiple responses to a single edge.

There are five steps to the Canny edge detection algorithm:

1. **Smoothing:** This is the first step of Canny algorithm, in which the input image is blurred using a Gaussian filter to remove noise. Figure 7.16a is the sample input image, used to show the steps of the Canny edge detector, Fig. 7.16b is a smoothed image generated by a Gaussian filter.
2. **Finding gradients:** The edges should be marked where the gradients of the image have large magnitudes. Gradients in x and y directions (G_x, G_y) are computed using a Sobel mask as above. The edge strength, i.e., gradient magnitude, is then calculated using Euclidean distance as follows:

$$|G| = \sqrt{G_x^2 + G_y^2} \qquad (7.5)$$

   Manhattan distance can also be used in order to reduce the computational complexity:

$$|G| = |G_x| + |G_y| \qquad (7.6)$$

   Gradient magnitudes $|G|$, can determine the edges quite clearly. However, the directions of the edges play an important role in detecting the broader edges more clearly. The directions of the edges are determined using:

$$\theta = \tan^{-1}\left(\frac{|G_y|}{|G_x|}\right) \qquad (7.7)$$

   The resultant gradient image, corresponding to the smoothed input image, is shown in Fig. 7.13c.
3. **Non-maximum suppression:** To convert the blurred edges of the gradient image (obtained in step 2) into sharp edges, this processing is performed. This is achieved by considering the edges only at the local maxima and discarding anything else (Fig. 7.13d).

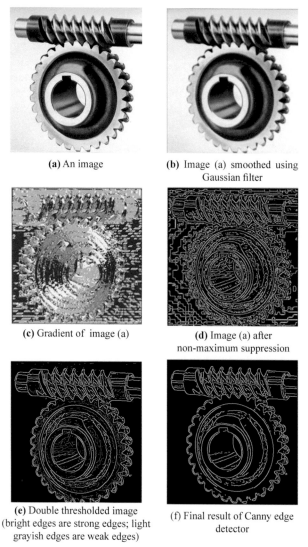

(a) An image

(b) Image (a) smoothed using Gaussian filter

(c) Gradient of image (a)

(d) Image (a) after non-maximum suppression

(e) Double thresholded image (bright edges are strong edges; light grayish edges are weak edges)

(f) Final result of Canny edge detector

Fig. 7.13. Results of various steps of Canny edge detector.
(For color images of Figure 7.13(c) see Color Figures Section at the end of the book)

4. **Double thresholding:** Most of the edges in the image, obtained from the non-maximum suppression step, are true edges present in that image. However, some edges may be caused by noise or color variations. Hence, the double thresholding mechanism is used in order to detect true edges:

Define two thresholds – L(lower threshold), H (higher threshold)
   If edge-pixel strength is less than L, not an edge
   If edge-pixel strength greater than H, strong edge
   If edge-pixel strength between L and H, weak edge.

Hence, all the edges stronger than the higher threshold would be marked as strong edges; on the other hand, edge pixels between the two thresholds are marked as weak, whereas pixels weaker than the lower threshold will be discarded. Figure 7.13e shows the image after double thresholding.

5. **Edge tracking by hysteresis:** All strong edges are considered final true edges. However, weak edges should be considered true edges only if these edges are connected with some strong edges, this is because some weak edges may be the result of noise or color variations. The basic idea is that weak edges, due to true edges, are much more likely to be connected directly to strong edges. This is done via the process of edge tracking. Edge tracking is implemented by BLOB-analysis (Binary Large OBject). The edge pixels are divided into connected BLOB's using an 8-connected neighborhood. BLOB's containing at least one strong edge pixel are then preserved, while the other BLOB's are suppressed. Figure 7.13f shows the final result of edged detection using the Canny edge detector. Comparative visual results of different gradient-based edge detectors are illustrated in Fig. 7.12. The results in Fig. 7.12e are generated from the Canny edge detector and are much better than the results of the Roberts (Fig. 7.12b), Prewitts (Fig. 7.12c) and Sobel (Fig. 7.12d) edge detectors.

The Canny edge detector can detect edges in noisy images with good accuracy because it first removes the noise by Gaussian filter. It generates one pixel wide ridges as the output by enhancing the signal with respect to the noise ratio by the non-maxima suppression method. Overall, it gives better edge detection accuracy by applying double thresholding and edge tracking hysteresis method. But the Canny edge detector is time-consuming and it is difficult to implement the Canny detector and reach real time response speeds.

7.4 Thresholding-based Image Segmentation

Thresholding is a commonly used method for image segmentation because of its simple implementation, low computation cost and intuitive properties. The thresholding-based image segmentation method divides pixels on the basis of intensity level or gray level of an image. Hence, this method divides the whole image into background and foreground regions where it is assumed that objects (foreground regions) have pixels of greater intensity levels as compared to the pixels in the background. Due to the aforementioned reason, thresholding-based methods are applicable where the objects in the image differ in their gray level distribution. In thresholding-based methods, all the pixels belonging to an object are given a value of "1" while background pixels are given value of "0". Therefore, a binary image (image with pixel values 0 or 1) is generated.

Furthermore, the thresholding-based segmentation method is roughly divided into two categories, global thresholding and variable thresholding, as described below:

7.4.1 Global Thresholding

As mentioned earlier, thresholding-based segmentation is used if the intensity distribution of objects and background pixels are sufficiently distinct in an image. The same idea is also applicable for global thresholding. In the global thresholding method, the same threshold value is used for the whole image. Global thresholding is used in many applications because of the simplicity and low computation cost of segmentation. Let, $f(x,y)$ and $g(x,y)$ be the input image and segmented image, respectively, then segmentation of the image is achieved as follows:

$$g(x,y) = \begin{cases} 1 & if f(x,y) > T \\ 0 & if f(x,y) \leq T \end{cases} \tag{7.8}$$

Accuracy of the segmentation depends on the appropriate selection of the threshold value T. Hence, deciding the value of T for an image is a challenge. Several methods for deciding the appropriate value of threshold T have been developed by researchers in recent years, such that the segmentation accuracy could be optimized. One simple method to select T is based on the intensity histogram of the input image. Figure 7.14b shows the intensity histogram of the input image (Fig. 7.14a). This histogram consists of two peak points; these two peaks correspond to the foreground (object) and background. Let P1 and P2 be the gray values of the peaks of the histogram. The threshold value T is considered as the mean value of P1 and P2:

$$T = \frac{P_1 + P_2}{2} \tag{7.9}$$

However, calculation of T as in Eq. 7.9 may not be appropriate when the variability in pixel values of the image is high as is the case in most of the applications. In practice, real images have a large variety of pixels. For this type of image, the threshold value T can be calculated using the following iterative algorithm:

1. Select an initial estimate (for example, average image intensity) for T.
2. Segment the image using T. This will produce two groups of pixels: G1 consisting of all pixels with gray level values > T and G2 consisting of pixels with gray level values <=T.

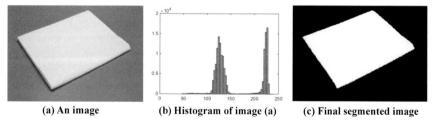

(a) An image   (b) Histogram of image (a)   (c) Final segmented image

Fig. 7.14. Result of global thresholding-based segmentation.

3. Find average gray levels m1 and m2 for the pixels in regions G1 and G2.
4. Calculate a new threshold value T = (m1+m2)/2.
5. Repeat steps 2 through 4 until the difference in Ts in successive iterations is smaller than a predefined parameter T0.

There are a number of other global thresholding techniques: Otsu, optimal thresholding, maximum correlation thresholding, clustering, Multispectral thresholding and Multithresholding.

The MATLAB code for image segmentation using a global thresholding-based method is similar to:

```
img = imread ('img.jpg');
figure, imshow(img); %show input image
histogram(img) %plot the histogram
Threshold_level = graythresh(img);
BW = im2bw(img, Threshold _level) %show segmented image
figure, imshow(BW);
```

7.4.2 Variable Thresholding

As discussed above, in global thresholding a single value of T is considered, however, if the value of threshold T varies over the image then it is known as variable thresholding. There are two types of variable thresholding methods, local thresholding and adaptive thresholding. In the local thresholding method, the value of T depends upon the neighborhood of the pixel being processed, whereas in the adaptive thresholding method the value of T is a function of pixel coordinates. Due to high computation complexity, adaptive thresholding is not suitable for real-time applications.

The advantages of thresholding-based segmentation methods are simple to implement and overall the segmentation process is quick and efficient for images with text.

Thresholding-based segmentation methods perform poorly when the images have lots of variation in the intensities of pixels.

7.5 Region Based Image Segmentation

A region in an image is defined as a group of connected pixels having similar properties. In general, objects also can be interpreted as a group of connected pixels of similar properties. Hence, regions can be used as a key for image segmentation. In region-based segmentation, pixels corresponding to an object are grouped together. Region-based segmentation techniques incorporate two basic similarity measures, these are value similarity (pixels similar to their neighbors are considered in the region) and spatial proximity (spatially close pixels are considered in the region). Hence, a region or object in any image consists of spatially close pixels with similar gray values. Two main techniques of region-based segmentation are region growing and region splitting and merging.

7.5.1 Region Growing

The region growing-based segmentation method is a bottom-up approach that starts from a set of initial pixels (called seeds). These seeds then grow on the basis of some similarity measures (gray level, color, texture). These seeds can be selected either manually (based on some prior knowledge) or automatically (depending on particular application).

A pixel is added to a region if it satisfies three conditions:

1. It has not been assigned to any other region.
2. It is a neighbor of that region.
3. The new region that was formed after the addition of the pixel is still uniform.

The criteria listed above can be described formally in order to form the complete region-growing algorithm. Let $f(x,y)$ be the input image; $S(x,y)$ is the binary image with the seeds (consists of value 1 only in the location of seeds, otherwise 0); Q is the predicate to be tested for each location (x,y) based on 8-connectivity region-growing algorithms:

1. Find all connected components in S and erode all the connected components of S until they are only one pixel wide:

$$S(x,y) = \begin{cases} 1 \ ; & \text{if seed is located at the pixel } (x,y) \\ 0 \ ; & \text{Otherwise} \end{cases}$$

2. Generate the binary image fQ:

$$fQ(x,y) = \begin{cases} 1 \ ; & \text{if } Q(x,y) \text{ is true} \\ 0 \ ; & \text{Otherwise} \end{cases}$$

3. Generate the binary image g such that $g(x,y) = 1$ if $fQ(x,y) = 1$ and (x,y) is 8-connected to a seed in S.
4. The resulting connected components in g correspond to the segmented regions.

There are two basic problems with region growing. Firstly, the accuracy of segmentation is dependent on suitable seed values; however, it is not a simple task to find good starting points. Secondly, in some images color may be the appropriate similarity criterion, while in others the image gray levels are a better choice.

7.5.2 Region Splitting and Merging (Quad Tree Method)

The main drawback of region growing is high computational cost. Because region growing starts from a very small seed it can be time consuming. To reduce the computational cost, the region split and merge method initially considers the whole image as a single region, then it repeatedly splits it into sub-regions until no more splits are possible. When the splitting is complete, two adjacent regions are merged on the basis of some similarity measure (such as graylevel, color, means, variance, texture, etc.). Merging is repeated until no further merging is possible.

Let f(x,y) be the input image and Q be the predicate to be tested. The region splitting and merging algorithm is as follows:

1. R1 = f
2. Subdivision in quadrants (Fig. 7.15) of each region Ri for which Q(Ri) = FALSE.
3. If Q(Ri) = TRUE for every regions, merge those adjacent regions Ri and Rj such that Q(Ri ∪ Rj) = TRUE; otherwise, repeat step 2.
4. Repeat step 3 until no further merging is possible.

Region-based image segmentation techniques are robust, since regions cover more pixels than edges and, thus, more information is available to characterize a region. The region could be detected based on some texture; however, this is not applicable in edge-based methods. Region-growing techniques are generally suitable in noisy images where edges are difficult to detect.

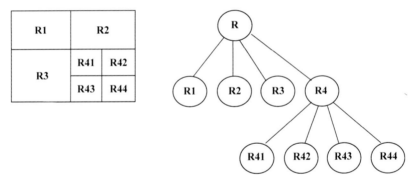

Fig. 7.15. Region splitting using Quad Tree.

7.6 Watershed Based Image Segmentation

In general, a watershed or drainage basin or catchment basin is defined as an area or ridge of land that separates waters flowing to different rivers, basins, or seas. Figure 7.16 illustrates catchment basin and watershed lines (lines dividing two or more catchment basins). This concept of watershed can be applied in image

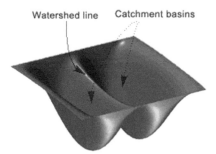

Fig. 7.16. Basic idea of watershed lines and catchment basins.
(For color image of this Figure see Color Figures Section at the end of the book)

processing to segment the image if the bright areas of the image are treated as high elevation (watershed lines) and the dark areas are assumed to be at a low elevation (catchment basin). Hence, all dark areas are analogous to watershed and bright areas are analogous to watershed lines. The process is as follows:

Consider an image I as a topographic surface and define the catchment basins and the watershed lines by means of a flooding process. Imagine that each local minima of the topographic surface S is pierced and then the water is released through the holes with constant speed into the watershed. During the flooding, two or more streams coming from different minima may merge. To avoid this merging, a dam has to be built on the merging points. In a similar manner, if releasing the water through the holes is prolonged, a dam is required in order to avoid the accidental merging of two watersheds. At the end of the process, only the watershed lines (boundaries of the dam) will be visible from the top view. These watershed lines separate the various catchment basins. Hence, the main objective of the watershed-based image segmentation is to identify the watershed lines.

Watershed transform-based segmentation performs better if foreground objects and background locations can be identified. The steps in the process are listed below.

Step 1: Convert the input image to grayscale.

Step 2: Obtain the gradient magnitude of the image.

Step 3: Identify the foreground objects in image.

Step 4: Compute the background markers.

Step 5: Compute the Watershed Transform of the Segmentation Function.

Step 6: Display the Result.

The output of various steps is shown in Fig. 7.17.

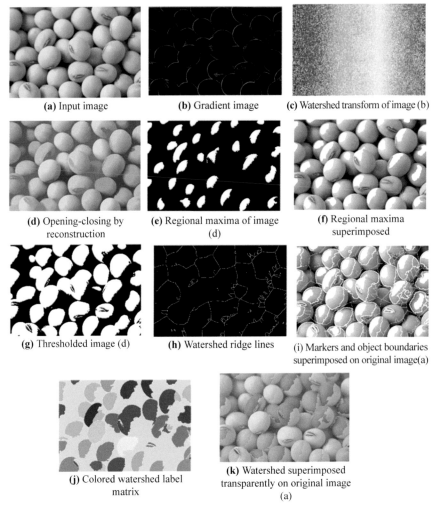

(a) Input image   **(b)** Gradient image   **(c)** Watershed transform of image (b)

(d) Opening-closing by reconstruction   **(e)** Regional maxima of image (d)   **(f)** Regional maxima superimposed

(g) Thresholded image (d)   **(h)** Watershed ridge lines   (i) Markers and object boundaries superimposed on original image(a)

(j) Colored watershed label matrix   **(k)** Watershed superimposed transparently on original image (a)

Fig. 7.17. Various steps of watershed-based image segmentation.
(For color images of Figure 7.17(c), (j), (k) see Color Figures Section at the end of the book)

Summary

- Image segmentation is an important process that is used to partition an image into various segments.
- Segmentation is an important step in mid-level and high-level image processing.
- Image segmentation techniques are based on continuity or discontinuity.
- The selection of an image segmentation technique depends on the contents of the image and the application for which the segmentation is performed.

8

Mathematical Morphology

8.1 Introduction

The word "morphology" comes from the Ancient Greek word *morphé*, which means "form", and *lógos*, which means "study, research". The term Morphology has been used in biology that deals with the study of the form and structure of organisms since the 17th century. The concept of Mathematical Morphology was given in the mid-sixties by Georges Matheron and Jean Serra of the Paris School of Mines in Fontainebleau, France. Then, in the 70s, it was used for image analysis research in the area of microscopic imaging in Europe. Publication of Serra's books on Image Analysis and Mathematical Morphology advanced the use of Mathematical Morphology in image processing research worldwide. More information on the birth of Mathematical Morphology can be found at http://cmm.ensmp.fr/~serra/pdf/birth_of_mm.pdf.

In digital image processing, mathematical morphology is used by applying some basic morphological operations that are used to investigate the interaction between an image and a certain chosen structuring element. These operations are non-linear in nature. Morphology is based on the assumption that an image consists of structures which can be processed by operations of set theory.

Morphological operations are primarily applied to binary images but can be extended to grayscale or color images. A grayscale image contains pixels having intensity ranging from 0 to 255, while a binary image has only two intensity values, viz. 0 and 1. A grayscale image can be converted to binary image by the process known as Binarisation.

In MATLAB®, a color image can be converted into grayscale image using:

$$I = \text{rgb2gray(RGB image)} \qquad (8.1)$$

The rgb2gray function converts a RGB image to grayscale image by eliminating the hue and saturation information while retaining the luminance. This function converts RGB values of an image to grayscale values by calculating a weighted sum of the R, G, and B components as:

$$0.2989 * R + 0.5870 * G + 0.1140 * B \tag{8.2}$$

The MATLAB function im2bw(I, level) converts the grayscale image I to a binary image. In the output image, all pixels in the input image with luminance value greater than level with the value 1 (white) are replaced by 1 and all other pixels are replaced by 0 (black). The value of level needs to be in the range [0,1]. A level value of 0.5 is midway between black and white, regardless of class. If the value of level is not specified, then the im2bw function considers the default value 0.5.

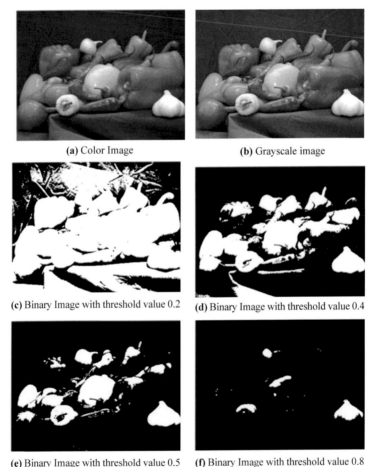

(a) Color Image   (b) Grayscale image

(c) Binary Image with threshold value 0.2   (d) Binary Image with threshold value 0.4

(e) Binary Image with threshold value 0.5   (f) Binary Image with threshold value 0.8

Fig. 8.1. (a) Color image (b) Corresponding grayscale image (c)–(f) Corresponding grayscale images converted into binary image using various threshold values.
Original image reprinted with permission of The MathWorks, Inc.
(For color image of Fig. 8.1(a), see Fig. 1.1(a) in Color Figures Section at the end of the book)

In Morphology, fundamental objects are sets and all the operations are based on the concepts of set theory. In the context of an image, a set is a group of pixels. The basic set theory operations are: union of two sets, intersection of two sets, difference of two sets, translation and complement of a set.

- **Union** of two images A and B, denoted as $A \cup B$, is an image containing pixel values which are either part of A or B or both.
- **Intersection** of two images A and B, denoted by $A \cap B$, is the image whose pixel values are common to both images A and B.
- **Difference** of two images A and B, denoted by A-B, also known as relative complement of B relative of A, subtracts from image A all elements which are in image B. A-B may not be same as B-A.
- **Complement** of an image is defined as the set consisting of everything other than the image.
- **Translation** shifts the origin of an image to some other point.

Some set operations are illustrated in Fig. 8.2.

The "union" set operation, $A \cup B$, is equivalent to the logical "OR" operation for binary images. The "intersection" set operation, $A \cap B$, is equivalent to the logical "AND" operation for binary images.

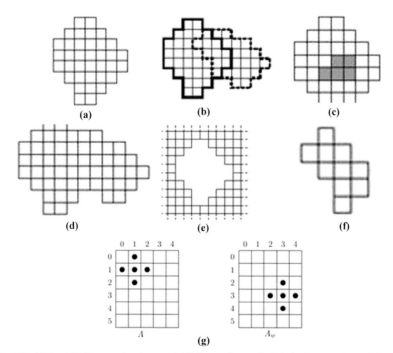

Fig. 8.2. (a) A set (b) Two overlapping sets (c) Sub-set of a set: shaded pixels are a subset of the set (d) Union of two sets (e) The complement of a set (f) The intersection of two sets A and B (g) Translation of an image.

8.2 Morphological Operations

Mathematical morphology has a wide range of operators which can be used in image processing. For mathematical morphology of binary images, the sets are considered as group of pixels in an image having value either 0 (black) or 1 (white).

In a binary image, an object is considered as a connected set of pixels. The connectivity can be either 4-connectivity or 8-connectivity (Fig. 8.3). In 4-connectivity, a pixel has four connected neighbors top, bottom, right and left. In 8-connectivity, the diagonally connected pixels are also considered, i.e., each pixel has eight connected neighbors including the diagonal pixels.

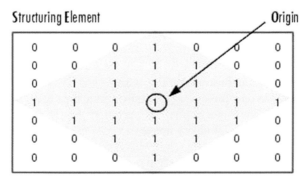

Fig. 8.3.   Origin of a structuring element using 8x8 neighborhood.

In morphological image processing, the operations are applied to an input image using a structuring element and an output image of the same size is generated. In a morphological operation, the pixel values of the output image are based on a comparison of a corresponding pixel in the input image with its neighbors. Two basic morphological image operations are dilation and erosion. The process of dilation adds pixels to the boundaries of objects in an image, while erosion is opposite to dilation and removes pixels on object boundaries. The number of pixels to be added or to be removed from the objects in an image depends on the size and shape of the structuring element which is used to process the image. The other morphological operations are obtained by combining these two operations.

8.2.1 Structuring Element

In morphological operations, the structuring element is very important. The morphological operations are controlled by a shape known as a structuring element. The structuring element consists of a pattern which is specified in terms of coordinates of a number of discrete points relative to some origin. A structuring element may be either flat or non-flat. In processing with a flat structuring element, a binary valued neighborhood is used, in which the true pixels are used in the morphological computation. A flat structuring element is actually a matrix of 0's and 1's, generally much smaller than the image being processed. The center pixel of the structuring element is called the origin and corresponds to the pixel being processed. The pixels in the structuring element containing 1's define the neighborhood of the structuring element. A non-flat structuring element is 3-dimensional; it uses 0's and 1's to define the extent of the structuring element in the x- and y-plane and the third dimension adds height value. The structuring element has to be selected based on the application. For example, to find lines in an image, a linear structuring element is selected. In a morphological operation, the origin of the structuring element is overlapped to each pixel position in the image and then the points within the translated structuring element are compared with the underlying image pixel values.

In a morphological function, the following code is used to find the origin of structuring elements:

$$\text{origin} = \text{floor}((\text{size}(\text{nhood})+1)/2) \tag{8.3}$$

where nhood is the neighborhood defining the structuring element.

8.2.2 Dilation

In binary images, dilation (also known as 'Minkowski Addition') is an operation that expands the size of foreground objects and reduces the size of holes in an image. In this process, structuring element B is moved over the image A and the intersection of B reflected and translated with A is the result of dilation. The pattern of growth depends on the structuring element used.

Dilation of an image A using structuring element B is represented as:

$$\text{Dilation } A \oplus B = \{z : (\widehat{B})_z \cap A \neq \emptyset\}, \text{ where } \emptyset \text{ is the empty set.} \tag{8.4}$$

The process of Dilation can be described as:

Initially, superimpose the structuring element B over every pixel of the image A. This is to be processed in such a way that the center of the structuring element coincides with the input pixel position.

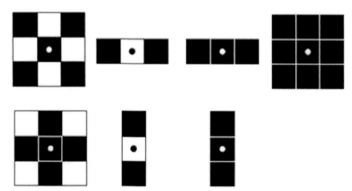

Fig. 8.4. Binary structuring elements (Black pixels are part of the element).

If at least one pixel of the structuring element B coincides with a foreground pixel in the image A, including the pixel being tested, then the output pixel in a new image is set to the foreground value.

Thus, some of the background pixels of the input image are converted into foreground pixels in the output image. Foreground pixels in the input image are not changed and these pixels remain foreground pixels in the output image also. In 8-connectivity, if a background pixel has at least one foreground neighbor then it becomes foreground; otherwise, it remains unchanged. The pixels which are changed from foreground to background are those pixels which lie at the edges of foreground regions in the input image; therefore, foreground regions grow in size and foreground features tend to connect or merge. Background features or

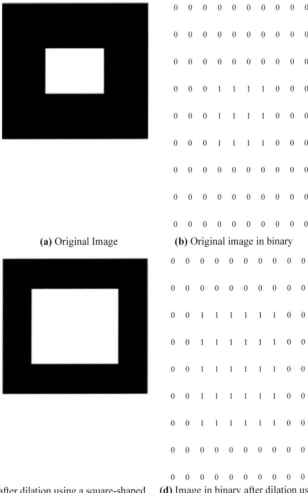

(a) Original Image

(b) Original image in binary

(c) Image after dilation using a square-shaped structuring element

(d) Image in binary after dilation using a square-shaped structuring element

Fig. 8.5 contd. ...

...*Fig. 8.5 contd.*

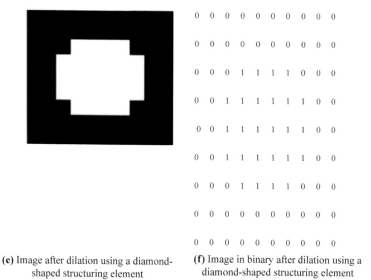

0	0	0	0	0	0	0	0	0	0
0	0	0	0	0	0	0	0	0	0
0	0	0	1	1	1	1	0	0	0
0	0	1	1	1	1	1	1	0	0
0	0	1	1	1	1	1	1	0	0
0	0	1	1	1	1	1	1	0	0
0	0	0	1	1	1	1	0	0	0
0	0	0	0	0	0	0	0	0	0
0	0	0	0	0	0	0	0	0	0

(e) Image after dilation using a diamond-shaped structuring element

(f) Image in binary after dilation using a diamond-shaped structuring element

Fig. 8.5. An image and the resultant image after dilation using various structural elements.

holes inside a foreground region shrink due to the growth of the foreground, and sharp corners are smoothed. Repeated dilation results in further growth of the foreground regions.

The structuring element can be considered analogous to a convolution mask, and the dilation process analogous to convolution, although dilation is based on set operations whereas convolution is based on arithmetic operations. After being reflected about its own origin it slides over an image, pushing out the boundaries of the image where it overlaps with the image by at least one element. This growing effect is similar to the smearing or blurring effect of an averaging mask. One of the basic applications of dilation is to bridge gaps and connect objects. Dilation with a 3x3 structuring element is able to bridge gaps of upto two pixels in length.

An important application of the dilation process is obtaining the boundaries of objects in an image. Dilation can be used to create the outline of features in an image (Fig. 8.6). If a binarized image is dilated once, and then the original image is subtracted pixel-by-pixel from the dilated image, the result is a one-pixel wide

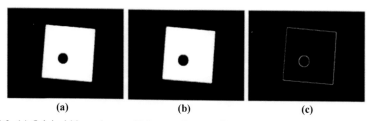

   (a)   **(b)**   **(c)**

Fig. 8.6. (a) Original binary image, (b) image after one dilation (c) result of subtracting image (a) from image (b).

boundary of the features in the original image. This operation tends to be more robust than most edge enhancement operations in the presence of image noise.

Dilation in MATLAB is performed using the following function:
    imdilate(image,kernel)

8.2.3 Erosion

Erosion, also known as 'Minkowski Subtraction', is a process that increases the size of background objects and shrinks the foreground objects in binary images. In the erosion of an image A by the structuring element B is the set of points z, such that B, translated by z, is contained in A.

Erosion of an image A by the structuring element B is represented as:

$A \ominus B = \{z: (B)_z \cap A \neq \emptyset\}$, where $\emptyset$ is the empty set.

In Erosion, the structuring element is superimposed onto each pixel of the input image, and if at least one pixel in the structuring element coincides with a background pixel in the image underneath, then the output pixel is set to the background value. Thus, some of the foreground pixels in the input image become background pixels in the output image; those that were background pixels in the input image remain background pixels in the output image. In the case of 8-connectivity, if a foreground pixel has at least one background neighbor then it is changed, otherwise it remains unchanged. The pixels which change from foreground to background are pixels at the edges of background regions in the input image, so the consequence is that background regions grow in size, and foreground features tend to disconnect or separate further. Opposite to the dilation, in erosion, background features or holes inside a foreground region grow, and corners are sharpened. Further erosion results in the growth of the background and shrinkage of the foreground.

Dilation is the dual of erosion, i.e., dilating foreground pixels is equivalent to eroding the background pixels.

Again, erosion can be considered analogous to convolution. As the structuring element moves inside the image, the boundaries of the image are moved inwards because image foreground pixels in the image are changed to background pixels wherever the structuring element overlaps the background region by at least one element.

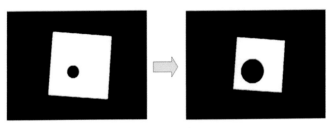

Fig. 8.7. Erosion.

Erosion can be used to eliminate irrelevant details, below a certain size, in an image. A structuring element eliminates detail in an image smaller than its own size. Erosion can be used to create a one-pixel wide outline of the features in an image by subtracting the eroded image from the original image.

Erosion is the dual of dilation, i.e., eroding foreground pixels is equivalent to dilating background pixels. However, erosion of an image, followed by dilation of the result, or vice versa, does not produce the original image; isolated foreground pixels removed during erosion, for example, are not re-instated during dilation.

Erosion can help in the counting of features which touch or overlap in an image.

It is also possible to perform a constrained or conditional dilation. In this process, an image, known as the seed image, is dilated but not allowed to dilate outside the mask image, i.e., the resulting features are never larger than the features in the mask image. This can be very useful in feature extraction and recognition.

Erosion in MATLAB is performed using the following function:

imerode(image,kernel)

8.2.4 Opening and Closing

Opening and closing are the two most widely used morphology operators that are defined in terms of combinations of erosion and dilation.

Opening is defined as erosion followed by dilation using the same structuring element for both operations. The erosion part removes some foreground pixels from the edges of regions of foreground pixels, while the dilation part adds foreground pixels. The foreground features remain roughly the same size, but their contours are smoother. As with erosion itself, narrow isthmuses are broken and thin protrusions are eliminated.

Opening of an Image A using a structuring element B can be defined as:

$$A \circ B = (A \ominus B) \oplus B \qquad (8.5)$$

Similar to erosions and dilations, the opening operator is based on the local comparison of a shape, the structural element, with the object that will be transformed. If the structural element is included in the object at the time of processing than the whole structural element will appear in the output of the transformation, otherwise not all the points of the structuring element will appear.

The effect of opening on a binary image depends on the shape of the structuring element. Opening preserves foreground regions that have a similar shape to the structuring element, or the structuring element completely contained in the region, while it tends to eliminate foreground regions of dissimilar shapes. Thus, binary opening can be used as a powerful shape detector for preserving certain shapes and eliminate others.

Figure 8.8a shows an image that contains various objects of shape lines and circles, with the diameter of the circles being greater than the width of the lines. If a circular structuring element with a diameter just smaller than the diameter of the

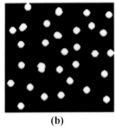

(a)   (b)

Fig. 8.8. (a) An image comprising both lines and circles, (b) the result after applying opening on (a) with a circular structuring element.

smallest circles is used to open this image, the resulting image Fig. 8.8b contains only the circles and the lines have been eliminated.

Opening can be visualized as:

Move the structuring element around inside each foreground object. Pixels covered by the structuring element entirely within the object are preserved. Foreground pixels which are not reached by the structuring element without it protruding outside the object are eroded away. When the structuring element is a circle, or a sphere in three dimensions, this operation is known as a rolling ball, and is useful for subtracting an uneven background from grayscale images.

Closing is defined as the dilation process followed by erosion, using the same structuring element for both operations. Closing smoothens the contours of foreground objects, merges narrow breaks or gaps and eliminates small holes.

Closing of an image A using a structuring element B is defined as:

$$A \cdot B = (A \oplus B) \ominus B \tag{8.6}$$

Similar to opening operator, the key mechanism of the closing operator is the local comparison of the structural element with the object to be transformed. If, at a given point, the structural element is included in the complement of the image, then the whole structural element will appear in the complement of the transformed image, otherwise the whole structural element will not appear in the output.

Closing can be used to eliminate the smaller holes in the image. Figure 8.9a shows an image containing holes of different sizes. A circular structural element of size between the diameter of the two sets of holes is used to close the image in the resulting image Fig. 8.9b. In Fig. 8.9b, the small holes are removed, leaving only the larger holes. This is because the larger holes allow the structuring element to move freely inside them without protruding outside.

Opening and closing operations are frequently used to remove artefacts in a segmented image before proceeding to further analysis. The choice of whether to use opening or closing, or a sequence of erosions and dilations, depends on the individual processing requirement. For example, if an image has foreground noise or if there is a need to eliminate long, thin features in an image, then opening is used. It should not be used in the situation where there is a possibility that the initial

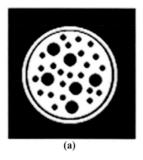 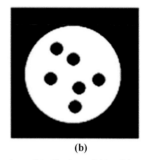

(a)   (b)

Fig. 8.9. (a) An image containing holes of two different sizes, (b) Closing of (a) with a circular structuring element of size between the two sets of holes.

erosion operation might disconnect regions. Closing is used if a region has become disconnected and there is a need to restore connectivity. If different regions are located close to each other, then closing may connect these regions; therefore, in this situation, closing should not be used.

Similar to erosion and dilation, opening and closing are also duals of each other, i.e., opening the foreground pixels with a particular structuring element is equivalent to closing the background pixels with the same structuring element. Opening and closing operations are idempotent, i.e., repeated application of either of them has no further effect on an image.

8.2.5 Hit-or-Miss Transform

The hit-or-miss transform is a basic tool that is used in mathematical morphology. The hit-or-miss transform uses a pair of structuring elements. This transformation is very useful in applications related to the matching of isolated foreground pixels or endpoint pixels of line-segments. The hit-or-miss transform is based on two erosions. In this transform, we search for points where B1 fits in an object, and B2 fits in the background.

Hit-or-miss transformation of an image A by structuring element pair B = (B1, B2) can be defined as:

$$A \otimes B = (A \ominus B_1) \cap (A^c \ominus B_2) \tag{8.7}$$

Almost all morphological operations can be derived from the hit-or-miss transformation.

An example of use of the hit-or-miss transform is in finding the corners in a binary image:

The hit-or-miss transform needs to be used four times with four different structuring elements representing the four right-angle corners found in binary images (Fig. 8.10), and then the four results are combined, using a logical "OR", in order to get the final result which shows the locations of all right-angled corners

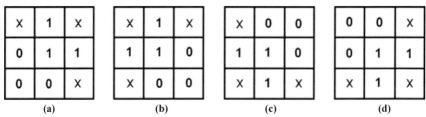

Fig. 8.10. Structuring elements used for finding corners using the hit-or-miss transform. (a) detects bottom left corners (b) detects bottom right corners (c) detects top right corners and (d) detects top left corners.

in any orientation. Figure 8.10 shows the final result of locating all the right-angle corners of a feature.

Different structuring elements can be used for locating other features within a binary image, isolated points in an image, or end-points and junction points in a binary skeleton, for example.

8.2.6 Thinning and Thickening

In some applications, the user may be interested not simply in whether or not an object is a single connected component, but in properties such as the number of branches and holes. Such properties are referred to as the topology of the object. Operations which preserves the topology of an object are called homeomorphisms. These operations are known as thinning operations and thickening operations.

Thinning is a morphological operation that repeatedly erodes away foreground pixels from the boundary of binary images while preserving the end points of line segments. Thickening is the dual of thinning, i.e., thickening the foreground is equivalent to thinning the background.

The thinning operation is related to the hit-and-miss transform and can be expressed quite simply in terms of it. The thinning of an image A by a structuring element B is:

thin $(A, B) = A - $ hitormiss (A,B), where the subtraction is a logical subtraction defined as $X - Y = X \cap $ NOT Y.   (8.8)

Figure 8.11 shows some examples of standard structuring elements used for thinning. The shaded squares must be contained in the object and the empty squares must lie outside the object. These structuring elements are used in the order 1, 2, ...8.

Iterated thinning until the image no longer changes results in a single-pixel wide skeleton or center line of an object in the image.

The thickening operation is the equivalent of thinning the complement of the set.

One use of skeletonization is to reduce the thresholded output of an edge detector, such as the Sobel operator, to a line measuring one pixel in thickness. Skeletonization is to be implemented as a two-step process that does not break the

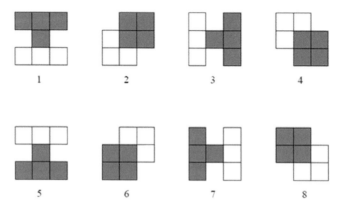

Fig. 8.11. Structuring elements for thinning.

objects. The first step is normal thinning, but it is conditional; that is, pixels are marked as candidates for removal, but are not actually eliminated. In the second pass, those candidates which can be removed without destroying connectivity are eliminated, while those that cannot are retained. The process is then repeated several times until no further change occurs, i.e., until convergence, thereby, the skeleton is obtained. Skeletonization preserves the topology, i.e., the extent and connectivity, of an object. The skeleton should be minimally eight-connected, i.e., the resulting line segments should always contain the minimal number of pixels that maintain eight-connectedness, and the approximate end-line locations should also be maintained.

8.2.7 Extension to Grayscale Images

Grayscale dilation and erosion are, thus, seen to be identical to convolution with the maximum and minimum rank masks, which operate like the median mask. The neighborhood around each pixel, and the pixels, are ordered by rank. If the center pixel is replaced by the maximum value in the neighborhood, grayscale dilation occurs. If the center pixel is replaced by the minimum value in the neighborhood, grayscale erosion occurs. If the center pixel is replaced by the median value in the neighborhood, median filtering occurs. Thus, dilation brightens and expands brighter areas of an image, and darkens and shrinks the darker areas. Erosion is the dual and has the opposite effect.

Grayscale opening and closing have the same form as their binary counterparts, i.e., grayscale opening is grayscale erosion followed by grayscale dilation, and grayscale closing is grayscale dilation followed by grayscale erosion.

Opening a grayscale image with a circular structuring element can be viewed as having the structuring roll under the profile of the image pushing up on the underside. The result of the opening process is the surface of the highest points reached by any part of this rolling ball. Conversely, grayscale erosion can be viewed as the rolling ball traversing the image profile and pressing down on it, with the result being the surface of the lowest points reached by any part of the rolling ball.

Non-linear processing is often used to remove noise without blurring the edges in the image. Recall how the median mask out-performed the linear averaging mask in removing salt-and-pepper noise. Morphological processing is often used due to its ability to distinguish objects based on their size, shape or contrast, viz. whether they are lighter or darker than the background. It can remove certain objects and leave others intact, making it more sophisticated at image interpretation than most other image processing tools.

Grayscale opening smoothens an image from above the brightness surface, while grayscale closing smoothens it from below. They remove small local maxima or minima without affecting the gray values of larger objects. Grayscale opening can be used to select and preserve particular intensity patterns while attenuating others.

A sequential combination of these two operations (open-close or close-open) is referred to as morphological smoothing can be used to remove 'salt-and-pepper' noise.

In images with a variable background, it is often difficult to separate features from the background. Adaptive processing is a possible solution. An alternative solution is so-called morphological thresholding, in which a morphologically smoothed image is used to produce an image of the variable background which can then be subtracted from the original image.

Morphological sharpening can be implemented by the morphological gradient, MG, operation

$$MG = \tfrac{1}{2} (Max (Image) - Min (Image)) \tag{8.9}$$

If a symmetrical structuring element is used, such sharpening is less dependent on edge directionality than sharpening masks such as the Sobel masks.

The morphological top hat transformation, TH, is defined as

$$TH = Image - Open (Image) \tag{8.10}$$

This is the analog of unsharp masking and is useful for enhancing detail in the presence of shading.

Granulometry is the name given to the determination of the size distribution of features within an image, particularly when they are predominantly circular in shape. Opening operations with structuring elements of increasing size can be used to construct a histogram of feature size, even when they overlap and are too cluttered to enable detection of individual features. The difference between the original image and its opening is calculated after each pass. At the end of the process, these differences are normalized and used to construct a histogram of feature-size distribution. The resulting histogram is called the pattern spectrum of the image.

Summary

- Mathematical morphology concerns the analysis of geometrical structures based on set theory, topology, etc., that can be applied in digital image processing.
- In a binary image, an object is considered as a connected (4-connected or 8-connected) set of pixels.
- Basically, morphological operations are defined for binary images.
- Two basic morphological operations are Dilation and Erosion. Other morphological operations can be derived from these two operations.
- Dilation causes objects in the image to dilate or grow in size and this depends on the structure of the structuring element used in the process.
- Erosion causes objects in the image to shrink and this also depends on the structure of structuring elements used in the process.
- Opening and closing are two commonly-used morphology operators that are defined in terms of combinations of erosion and dilation.
- Opening is the erosion followed by dilation using the same structuring element.
- Closing is erosion of the dilation of that set using same structuring elements.

9

◇◇

Image Understanding

9.1 Introduction

After an image is segmented into various parts, each segmented part has to be represented and described in such a way that further processing can be performed on these parts. As the objects or parts of an image are a collection of pixels, there is a need to describe the properties of these pixels for recognition of these objects. Defining the shape of an object is a very complex and difficult task. The description of these objects is a set of numbers that can be used for matching and recognition. This set of numbers is called a descriptor and can be used for matching with descriptors of known objects to identify.

A descriptor should satisfy the following requirements:

a) Descriptors of two different objects should be different. Descriptors of two objects should be the same only if the objects are identical.
b) Descriptor should be invariant, e.g., a rotation invariant descriptor will not change if the object is rotated. An affine and perspective invariant descriptor can describe an object observed from various perspectives.
c) Descriptor should be congruent.
d) A descriptor should not contain redundant information. It should contain only the information required in order to describe the characteristics of an object. It should be a compact set.

A region can be represented using external representation or internal representation. The external representation uses shape characteristics, while the internal representation uses regional properties to describe an image.

9.2 Contour-Based Shape Representation and Description

A point is said to be on the boundary of an object or region if it is part of the object and there exists at least one pixel in its neighborhood that is not part of the object or region. To identify the boundary of an object, a point is identified on the boundary or contour, then, by moving around the contour either in clockwise, or anti-clockwise direction to the nearest connected contour point, the rest of the boundary is identified.

In MATLAB®, bwlabel function is used to find all the connected components in a binary image. Another useful function in MATLAB is bwperim, this returns a binary image with all boundary pixels of all the regions in the input image.

9.2.1 Chain Code

Chain code is one of the early techniques in image processing, suggested in the 1961 by Freeman. In chain codes, an object is described by a sequence of small unit-size line segments with a set of defined orientations.

Given a complete boundary, i.e., a set of connected components, then starting from one point of a segment, according to 4-connectivity or 8-connectivity, the direction of the next segment is determined (Fig. 9.1). The direction from one unit-size segment to next unit-size segment becomes an element of chain code. This process is repeated until the starting point is reached. Chain code of the object boundary is formed by concatenating these numbers.

The disadvantage of this code is that the code will be different for different starting points. Starting point invariance can be achieved by a variant of chain code using the following process: Obtain chain code to constitute an integer from the digits. Then, shift the digits cyclically (replacing the least significant digit with the most significant one, and shifting all other digits one place to the left) to make the smallest integer. This smallest integer code will be a code that is invariant to the starting point.

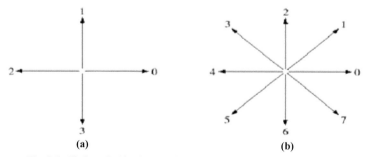

Fig. 9.1. Chain code (a) using 4-neighborhood (b) using 8-neighborhood.

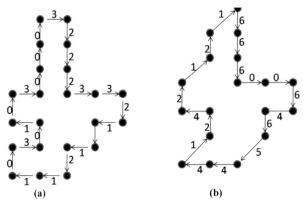

Fig. 9.2. Chain code of a boundary using (a) 4-directional chain code (b) 8-directional chain code.

For example:

If obtained chain code is 6553320000, then after performing cyclic rotation of digits, the smallest integer chain code will be 0000655332, which is invariant to the starting point.

- **Polygonal Approximation**

  An object boundary can be approximated by a polygon so that the object can be represented easily. The main aim of polygonal approximation is to approximate the boundary shape with the minimum possible polygonal segments. A number of polygonal approximation techniques are available with various levels of complexity.

- **Minimum Perimeter Polygons**

  Minimum perimeter polygons can be thought of as two walls; one for the outside boundary and the other for the inside boundaries of the strip of cells, and the object boundary as a rubber band contained within the walls. If the rubber band is allowed to shrink, it takes the shape of a polygon of minimum perimeter as shown in Fig. 9.3. The accuracy of a minimum perimeter polygon is determined by the size of the cells.

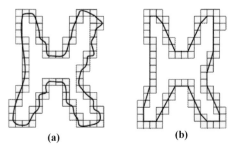

Fig. 9.3. (a) Object boundary (b) Corresponding Minimum perimeter polygon.

- **Merging Techniques**

  Polygonal approximation, based on average error or other criteria, is used to merge points along a boundary until the least square error line fit of the points merged so far exceeds a pre-set threshold. If it exceeds, the parameters of line are stored, and the error is set to 0. This process is repeated, merging new points along the boundary until the least square error again exceeds the threshold. Finally, the intersection of adjacent line segments form the vertices of approximated polygon (Fig. 9.4).

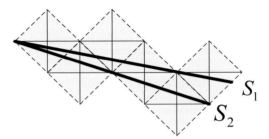

Fig. 9.4. Merging process.

9.2.2 Signature

Signature is a technique that translates a two-dimensional function (image) to one that is 1-dimensional. A simple technique to get the signature is to plot the distance from the center to the boundary as a function of angle (Fig. 9.5). Signatures are noise sensitive. To reduce noise sensitivity, contours may be smoothed before finding the signature.

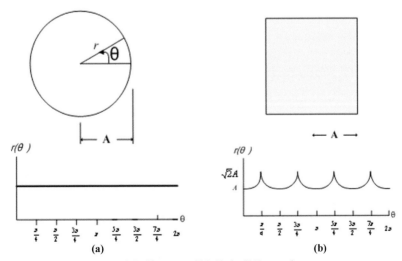

Fig. 9.5. Signature of (a) Circle (b) Rectangle.

9.2.3 Fourier Descriptors

Cosgriff, in 1960, suggested the use of Fourier Descriptors as a representation for closed curves. In this representation, a curve is drawn as tangential orientation against arc length. The resulting one-dimensional boundary profile is normalized to a length of 2, then it is described in the form of Fourier Series using the Fourier expansion:

$$c_n = \frac{1}{N} \sum_{s=0}^{N} \psi(x) e^{-j2\pi ns}$$

The boundary of any object can be represented using an infinite series of Fourier coefficients, c_n. In this series, the lower order terms denote the macroscopic behavior of the boundary curve and the higher order terms correspond to the curve's detailed microscopic behavior. The number of terms can be decided in order to make the descriptor finite. Fourier descriptors are rotationally invariant and scale invariant but not translationally invariant. They do not possess translational invariance and the respective shapes need to be located by segmentation. Fourier descriptors are very sensitive to boundary points, therefore, their performance degrades in the presence of noise or any distortion.

9.3 Boundary Segments Description

Decomposing a boundary into segments is useful in many applications as it reduces the boundary's complexity. If the segment type is known for all segments, the boundary of an object can be represented as a chain of these segment types. If the boundary contains one or more concavities, then this approach is very useful. Objects can be represented using planar graphs with nodes representing sub-regions generated from the decomposition of regions. The graph representation of the boundaries is easily understandable and insensitive to minor changes in shapes. These are also invariant with respect to translation and rotation.

9.3.1 Simple Shape Descriptors

Many simple shape descriptors based on region properties are available. Some of these are:

- **Area:** Area of an object is defined as the number of pixels in a particular region. It is very useful in the applications where relative size of various objects is important. In a rectangular raster, simply counting the region pixels gives the value of area. In certain cases, the actual area of covered pixels can be taken. If a region is represented by n polygon vertices (i_k, j_k) *and* $(i_0, j_0) = (i_n, j_n)$, then the area is defined as

$$area = \frac{1}{2} \left| \sum_{k=0}^{n-1} (i_k j_{k+1} - i_{k+1} j_k) \right|$$

- **Eccentricity**

  Eccentricity can be used to describe the area of an object. Many eccentricity measures are used for description, e.g., ratio of main region axes of inertia. Another is the ratio of the length of the maximum chord and the maximum chord perpendicular to it.

- **Euler Number**

  Euler number E is defined as:

  E = Number of contiguous parts of an object - number of holes in the object

  Euler number is topologically invariant.

- **Compactness**

  Compactness is independent of linear transformations. A circle is most compact region. It is defined as:

  $$compactness = \frac{(region\ border\ length)^2}{area}$$

  Some other scalar region descriptors are Projections, Direction, Rectangularity, Elongatedness.

9.3.2 Skeletons

Skeletons represents the structural shape of a region in the form of a graph. The skeleton of an object can be obtained by the iterations of thinning process. The skeleton of a region is defined through MAT (medial axis transformation), given by Blum in 1967.

In MATLAB, the bwmorph function is used to find the skeleton of an image.

9.3.3 Moments

Invariant moments (IM) are also called geometric moment invariants. Geometric moments, are the simplest of the moment functions with basis $\Psi_{pq} = x^p y^q$. Geometric moment function m_{pq} of order $(p+q)$ is:

$m_{pq} = \sum_X \sum_Y x^p y^q\ f(x,y),\ p,q - 0,1,2....$

The geometric central moments, which are invariant to translation, are defined as

$$\mu_{pq} = \sum_X \sum_Y (x-x')^p (y-y')^q\ f(x,y)\ with\ p,q = 0,1,2,\mathrm{K}\ .$$

where $x' = m_{10}/m_{00}$ $_{and}$ $y' = m_{01}/m_{00}$ and

A set of 7 invariant moments (IM) is:

$\phi_1 = \eta_{20} + \eta_{02}$

$\phi_2 = (\eta_{20} - \eta_{02})^2 + 4\eta_{11}^2$

$$\phi_3 = (\eta_{30} - 3\eta_{12})^2 + (\eta_{03} - 3\eta_{21})^2$$

$$\phi_4 = (\eta_{30} + \eta_{12})^2 + (\eta_{03} + \eta_{21})^2$$

$$\phi_5 = (\eta_{30} - 3\eta_{12})(\eta_{30} + \eta_{12})[(\eta_{30} + \eta_{12})^2 - 3(\eta_{21} + \eta_{03})^2] +$$
$$(3\eta_{21} - \eta_{03})(\eta_{21} + \eta_{03}) \times [3(\eta_{30} + \eta_{12})^2 - (\eta_{21} + \eta_{03})^2]$$

$$\phi_6 = (\eta_{20} - \eta_{02})[(\eta_{30} + \eta_{12})^2 - (\eta_{21} + \eta_{03})^2] + 4\eta_{11}^2(\eta_{30} + \eta_{12})(\eta_{21} + \eta_{03})$$

$$\phi_7 = (3\eta_{21} - \eta_{03})(\eta_{30} + \eta_{12})[(\eta_{30} + \eta_{12})^2 - 3(\eta_{21} + \eta_{03})^2] +$$
$$(3\eta_{12} - \eta_{03})(\eta_{21} + \eta_{03}) \times [3(\eta_{30} + \eta_{12})^2 - (\eta_{21} + \eta_{03})^2]$$

where $\eta_{pq} = \mu_{pq}/\mu_{00}^\gamma$ and $\gamma = 1 + \dfrac{p+q}{2}$ *for* $p + q = 2,3, \dots$

These are computationally simple and are invariant to scaling, rotation and translation. These moments are very sensitive to noise.

Besides the moments discussed above, there are many other moments for shape representation, for example, Zernike moments, radial Chebyshev moments (RCM), homocentric polar-radius moments, orthogonal Fourier-Mellin moments (OFMMs), pseudo-Zernike moments, etc. In general, the moment-based shape descriptors are concise, robust and easy to compute. They are also invariant to scaling, rotation and translation of the object. However, because of their global nature, the disadvantage of moment-based methods is that it is difficult to correlate high order moments with a shape's salient features.

9.4 Object Recognition

Object recognition is actually pattern classification in which a class is assigned to an image (or objects in the image) based on a numerical representation of the image (or object) properties. The features of the objects are identified and then extracted. These features are then classified using a classifier which determines the classes of the pattern.

9.4.1 Decision Function

When the number of classes is known and when the training patterns are such that there is geometrical separation between the classes, a set of decision functions is used to classify an unknown pattern. For example, if there are two classes, Class 1 and Class 2, which exist in R over n, and a hyperplane $d(x) = 0$ separates the two, then decision function $d(x)$ as a linear classifier can be used, classifying the new pattern as

if $d(x) > 0 \Rightarrow$ x belongs to Class 1

and if $d(x) < 0 \Rightarrow$ x belongs to Class 2

The hyperplane $d(x) = 0$ is called a decision boundary. When a set of hyperplanes can separate between m given classes in R over n, these classes are linearly separable. Sometimes, elements cannot be classified by using linear decision functions, then non-linear classifiers are used.

9.4.2 Statistical Approach

The training patterns of various classes often overlap, when they are originated by some statistical distributions, for example. In this case, a statistical approach is needed, especially when the various distribution functions of the classes are known. A statistical classifier must also evaluate the risk associated with every classification which measures the probability of misclassification.

The Bayes classifier, based on Bayes formula from probability theory, minimizes the total expected risk. In order to use the Bayes classifier, the pattern distribution function for each class must be known. If these distributions are unknown, then these distributions can be approximated using the training patterns.

9.4.3 Fuzzy Classifiers

Sometimes classification is performed with some degree of uncertainty. Either the classification outcome itself may be in doubt, or the classified pattern x may belong in some degree to more than one class. In such cases, fuzzy classification is used, where a pattern is a member of every class with some degree of membership between 0 and 1.

9.4.4 Syntactic Approach

Unlike the previous approaches, the syntactic pattern recognition utilizes the structure of the patterns. Instead of carrying an analysis based strictly on quantitative characteristics of the pattern, we emphasize the interrelationships between the primitives, the components which compare the pattern. Patterns that are usually subject to syntactic pattern recognition are character recognition, fingerprint recognition, chromosome recognition, etc.

9.4.5 Neural Networks

The neural network approach requires a set of training patters and their correct classification. The architecture of the neural net includes an input layer, output layer and hidden layers. It is characterized by a set of weights and an activation function which determine how any information (input data) is being transmitted to the output layer. The neural network is trained by training patterns and adjusts the weights until the correct classifications are obtained. It is then used to classify arbitrary unknown patterns.

Summary

- Shape representation and description is used to describe an object in computer understandable form.
- A number of shape description techniques are available, the selection of a particular technique depends on the application requirement.
- Chain code describes a region as a sequence of numbers called Freeman code.
- Geometric contour representations use geometric properties of described regions, e.g., boundary length, signature, etc.
- Some simple region-based descriptors are used to describe regions, e.g., Area, Eccentricity, Euler number.
- Skeleton can be used to describe an object in the form of a graph.
- Invariant moments (geometric moment invariants) are also used to describe any object.
- Object recognition is pattern classification in which a class is assigned to an image (or objects in the image) based on a numerical representation of the image in order to identify the object.
- A number of object recognition approaches are available, which can be used as per the requirement.

10

Image Compression

10.1 Introduction

In today's digital world, a huge amount of information in the form of images and videos is available. These images and videos are stored and transmitted from one device to another. Storage and transmission both require file size to be as small as possible. Data compression is a technique that is used to minimize the overall cost of data storage and transmission by reducing the size of a file. Image compression is an aspect of data compression, which reduces the amount of data required to represent a digital image without loss of generality.

In recent years, the storage and transmission facilities have grown rapidly, and a large amount of digital image data can be stored and transmitted easily. However, with the advancement of storage capacity and transmission bandwidth, digital image data is also growing exponentially. Thanks to the new High Definition (HD) technology being used in the field of digital images. For example, a grayscale 8-bit image of 512×512 pixels, contains 256 kilobytes data. If color information is also added, then the data size increases three times, and HD technology further increases the demand of storage to gigabytes. To store a video having 25 frames per second, even one second of color film requires approximately 19 megabytes of memory, thus, a typical storage medium of size 512 MB can store only about 26 seconds of film. This storage requirement of image or video files is very large, therefore, the file size needs to be reduced using compression.

10.2 History of Compression Technologies

Data compression began in the field of telegraphy in the early eighteenth century when Morse code was invented in 1838. In Morse code, the most frequently used letters of the English language, e.g., "e" and "t", are encoded using shorter codes. After a decade, with the advancement of information theory in 1940, many data compression techniques were developed. Later, in 1947, with the birth of mainframe computers, Claude Shannon and Robert Fano invented a coding technique known as Shannon-Fano coding. This algorithm assigns codes to symbols in a given block of data according to the probability of the symbol occurring. According to this algorithm, the probability of occurrence of a symbol is inversely proportional to the length of the code, therefore the code to represent the data is shorter.

In 1951, David Huffman invented a similar yet more efficient technique in which a probability tree is formed bottom-up. Later, in 1977, Abraham Lempel and Jacob Ziv invented an algorithm, known as LZW algorithm, that uses a dictionary to compress data. With the advent of digital image in the late eighties, many new image compression techniques evolved, such as JPEG.

10.3 Image File Types

The primary requirement of image compression is to identify the type of image that has to be compressed. From the compression point of view, the images types can be:

- Binary images
- Grayscale images
- Color images
- Video

Out of these types, grayscale, color images and videos are similar to each other in terms of compression algorithms. The techniques designed for grayscale images can also be applied efficiently to color images and videos, but the same argument is mostly untrue for binary images.

The images are stored in a system using different image formats. An image file format is a standardized way of organizing and storing digital images. An image file format may store data in uncompressed or compressed form. Some of the common image formats are:

- **Joint Photographic Experts Group (JPEG):** JPEG is a very common compressed image file format that can store 24-bit photographic images, i.e., an image having upto 16 million colors, such as those used for imaging and multimedia applications. JPEG compressed images are usually stored in the JPEG File Interchange Format (JFIF) due to the difficulty of programming encoders and decoders that fully implement all aspects of the standard. Apart from JFIF, Exchangeable image file format (Exif) and ICC color profiles have also been proposed in recent years to address certain issues related to JPEG standard. JPEG standard was designed to compress, color or grayscale continuous-tone images.

- **Graphics Interchange Format (GIF)**: GIF is a bitmap lossless image file format developed by Steve Wilhite in 1987 and commonly used in the World Wide Web due to its portability. GIF format is useful for black and white, grayscale images and color images having less than 256 colors. This format also supports animations and allows a separate palette of upto 256 colors for each frame. This format is well suited to simpler images such as graphics or logos with solid areas of color but, due to palette limitations, not very suitable for color photographs and other images with color gradients. Most color images are 24 bits per pixel and hence can't be stored as GIF. To store such images in GIF format, the image must first be converted into an 8-bit image. Although GIF format is a lossless format for color images with less than 256 colors, a rich truecolor image may "lose" 99.998% of the colors.

- **Portable Network Graphics (PNG):** PNG image compression format was designed as a replacement for Graphics Interchange Format (GIF), as a measure to avoid infringement of patent on the LZW compression technique. This format is lossless and supports 8-bit palette images (with optional transparency for all palette colors) and 24-bit true-color (16 million colors) or 48-bit true-color with and without alpha channel. In general, an image in a PNG file format can be 10% to 30% more compressed in comparison to a GIF format. This format maintains a trade-off between file size and image quality when the image is compressed.

  In comparison to JPEG, PNG format is better when the image has large, uniformly colored areas. PNG format also supports partial transparency that is useful in a number of applications, such as fades and anti-aliasing for text. PNG is a good choice for web browsers and can be fully streamed with a progressive display option.

- **Tagged Image File Format (TIFF):** TIFF is a flexible image file format that supports both lossless or lossy image compression. TIFF specification given by Aldus Corporation in 1986 normally saves eight bits per color (red, green, blue) for 24-bit images, and sixteen bits per color for 48-bit images. For handling images and data within a single file, TIFF file format has the header tags (size, definition, image-data arrangement, applied image compression) defining the image's geometry. Thus, the details of the image storage algorithm are included as part of the file. TIFF format is widely accepted as a photograph file standard in the printing business, scanning, fax machines, word processing, optical character recognition, image manipulation and page-layout applications. TIFF files are not suitable for display on web due to their large size and most web browsers do not support these files.

- **Bitmap (BMP):** BMP file format by Microsoft is an uncompressed, simple, widely accepted format.

10.4 Compression Quality Measures

A vast range of image compression techniques are available with a common aim of reducing the data required to represent an image. Usually, the efficiency of a compression technique is measured using compression amount or compression efficiency, algorithm complexity and distortion amount.

10.4.1 Compression Efficiency

Compression efficiency is measured in terms of the average number of bits required per stored pixel of the image, known as bit rate:

$$bit\ rate = \frac{Size\ of\ the\ compressed\ file}{Number\ of\ pixels\ in\ the\ image} = \frac{N}{K}(\text{bits per pixel}) \tag{10.1}$$

In case of very low bit rate, the compression ratio measure can be used. The compression ratio (C_R) is defined as the ratio between the size of an uncompressed image (N_1) and size of compressed image (N_2) as:

$$C_R = \frac{N_1}{N_2} \tag{10.2}$$

10.4.2 Algorithm Complexity

The complexity of a compression algorithm is measured in terms of both time and space. A compression algorithm is of no use in real time, if its processing time is too high. Therefore, fast decompression is desired in such applications. Similarly, space complexity helps to find out the maximum storage space required to run a particular compression algorithm.

10.4.3 Distortion

For measuring the distortion amount, either a subjective assessment or a quantitative assessment is used. In subjective criteria, the image quality assessment depends on the consensus of the individual end user. Both expert and non-expert viewers are equally involved in the experiment, and rate the compressed image based on some defined rating scale. This process is not practical and based on subjectivity.

On the other hand, quantitative assessment techniques, also known as objective analysis, are based on the mathematical modelling of both uncompressed as well as compressed images. In objective analysis, the original image is considered as perfect and all changes are considered as occurrences of distortion, irrespective of their appearance. The most commonly used measures for assessing image distortion due to lossy compression are the Mean Absolute Error (MAE), Mean Square Error (MSE) and Signal to Noise Ratio (SNR). Mean Absolute Error can be defined as:

$$MAE = \frac{1}{MN} \sum_{m=1}^{M} \sum_{n=1}^{N} \left| P(m,n) - \hat{P}(m,n) \right| \tag{10.3}$$

Here, $P(m, n)$ represents the original image and $\hat{P}(m, n)$ is the compressed image. Similar to MAE, another measure used to assess distortion is Mean Square Error (MSE). It can be defined as:

$$MSE = \frac{1}{MN} \sum_{m=1}^{M} \sum_{n=1}^{N} \left(P(m,n) - \hat{P}(m,n) \right)^2 \tag{10.4}$$

MSE is the most popular distortion measure. However, the notion of minimization of mean square error often does not agree with perceptual quality because it does not guarantee optimal perceptual quality. As an alternative, MAE is used as the quality measure by other studies. Similar to MSE, another distortion measure is Root Mean Square Error (RMSE), which is square root of MSE and gives more emphasis to the larger errors in the image.

Other than MSE, SNR is another popular distortion measure. It is defined as:

$$SNR = 10 \log_{10} \left(\frac{\sigma^2}{MSE} \right) dB \tag{10.5}$$

where, σ^2 represents the variance of the image having $M \times N$ pixels. It is defined as:

$$\sigma^2 = \frac{1}{MN} \sum_{m=1}^{M} \sum_{n=1}^{N} (P(m,n) - \mu)^2, \quad \mu = \frac{1}{MN} \sum_{m=1}^{M} \sum_{n=1}^{N} P(m,n) \tag{10.6}$$

where, $P(m, n)$ represents the gray value of the pixel at m^{th} row and n^{th} column, and μ is the mean of the image gray values. SNR is also used to derive yet another popular quantitative distortion measure known as peak SNR. It is defined as the ratio of square of the peak signal to the MSE:

$$SNR = 10 \log_{10} \left(\frac{peak^2}{MSE} \right) dB \tag{10.7}$$

PSNR is not well correlated with the visual quality of the image. In other words, two images having the same PSNR value may not have same visual quality. Consequently, other objective measures are also used in order to compute the distortion amount in the resultant image.

10.5 Image Redundancy

Image Compression techniques work on the principle of removing any redundancy present in an image. The amount of redundancy may vary from image to image. In Fig. 10.1, the image on the left has more redundancy in comparison to image on the right side as the image pixels in the image on the left have similar values to their neighbors, while in the image on the right, image pixel values are less consistent.

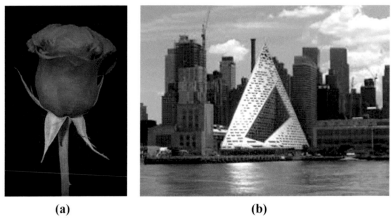

          (a)   **(b)**

Fig. 10.1. Two images with different amounts of redundancies.

The redundancy in an image can be of following types:

- Inter-Pixel Redundancy
- Psychovisual Redundancy
- Coding Redundancy

10.5.1 Inter-Pixel Redundancy

This type of redundancy mainly occurs due to statistical dependencies among image pixels, especially between neighboring pixels. In real world images, the neighboring pixels that belong to the same object or background tend to have similar intensity values except at boundaries or illumination changes. This correlation among the neighboring pixels results from the structural or geometrical relationship between the objects in the image. Due to the correlation, much of the visual information that many pixels carry is redundant. Information at a pixel can be easily predicted from their neighboring pixels quite accurately, thus resulting in redundancy reduction. Consequently, many compression techniques exploit inter-pixel dependency for image compression. Some examples are Run Length Encoding (RLE), predictive coding such as Differential Pulse Code Modulation (DPCM) and Constant Area Coding (CAC).

10.5.2 Psychovisual Redundancy

This type of redundancy arises due to the fact that human vision shows different sensitivity to different visual information. In other words, some visual information is less important for perception of an image than other visual information. This information is called psychovisual redundant information. For example, the human vision system is more adept at responding to rapid changes of illumination than it is at perceiving fine details and gradual changes. Moreover, human eyes are more sensitive to changes in brightness than to changes in color. The pixel values that exhibit the aforementioned information lead to psychovisual redundancy because the human brain does not perform the quantitative analysis of every pixel value in the image. Hence, in the process of compression, such psychovisual redundant information can be easily removed without compromising the image perception quality. Even though the loss is negligible psychovisually, the elimination results in loss of some quantitative information that cannot be recovered later. Hence, the compression techniques that take advantage of psychovisual redundancy are lossy compression techniques. Most of the image coding algorithms exploit this type of redundancy, such as the Discrete Cosine Transform (DCT)-based algorithm.

10.5.3 Coding Redundancy

This redundancy is due to inefficient data representation. In an uncompressed image, all the gray values are encoded using fixed length bit string. For example, an 8-bit binary string is used to represent 256 graylevels of a grayscale image, though the image may not be using all gray levels with equal number of occurrences. This leads to coding redundancy since the number of bits per pixel required to represent an image is more than the required number of bits. To reduce coding redundancy, variable length code can be used, i.e., shorter code words for frequent graylevels and longer code words for less common graylevels can be used, thus resulting in a reduction of the overall code length. Some popular compression techniques that exploit the coding redundancy are Huffman coding and Arithmetic Coding.

10.6 Fundamental Building Blocks of Image Compression

Generally, the main aim of any image compression technique is to reduce the quantity of data required to represent an image, without compromising the quality of the original image. However, minor loss in image quality is sometimes acceptable in certain applications, at the cost of amount of data reduced which is required to represent an image (Fig. 10.2). Based on different requirements, the image compression techniques can be broadly classified into two different classes:

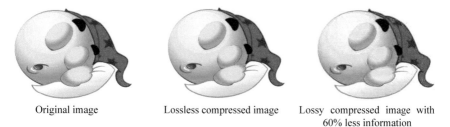

| Original image | Lossless compressed image | Lossy compressed image with 60% less information |

Fig. 10.2. Lossless and Lossy image compression.

- Lossless Compression Techniques
- Lossy Compression Techniques

In lossless compression techniques, the original image can be recovered perfectly, whereas in lossy techniques, the recovered image is not identical to the original one.

10.6.1 Lossless Image Compression

In a lossless compression algorithm, there is no loss of information after compression. These techniques are reversible and information preserving. The decompressed image is reconstructed without errors, i.e., no information is lost. Application domains include compression of text files, fax, technical drawing and digital archiving. In medical applications, for example, information loss is not accepted. Apart from information preservation, another reason to use lossless coding schemes is their lower computational requirements.

A lossless compression process has the following three stages:

- The model: the data to be compressed is analyzed with respect to its structure and the relative frequency of the occurring symbols.
- The encoder: produces a compressed bit-stream/file using the information provided by the model.
- The adaptor: uses information extracted from the data (usually during encoding) in order to continuously adapt the model (more or less) to the data.

The most fundamental idea in lossless compression is the use of shorter code-words (in terms of their binary representation) to replace frequently occurring symbols.

Each process which generates information can be thought of as a source emitting a sequence of symbols chosen from finite symbols.

Let S be the "source of information"; $\{S_1, S_2,\dots \dots \dots S_n\}$ are corresponding alphabets with occurrence probabilities $\{P(S_1), P(S_2),\dots \dots \dots P(S_n)\}$. Symbols are assumed to be independent, and the information provided by such symbols is the sum of information as given by the single independent symbols. The information provided by a symbol S_i related to its probability $P(S_i)$ is represented as:

$$I(S_i) = \log \frac{1}{P(S_i)} \tag{10.8}$$

The entropy is obtained which shows the average information per symbol:

$$H(S) = \sum_{i=1}^{n} P(S_i) I(S_i)$$

$$H(S) = -\sum_{i=1}^{n} P(S_i) \log_2 P(S_i) \, bits / symbol \tag{10.9}$$

In lossless compression, entropy helps us measure the number of bits required on average per symbol in order to represent the source. Entropy is a lower bound on the number of bits required to represent the source information.

- **Huffman Coding**

  Huffman Coding is an entropy encoding technique that uses variable length coding to form a tree, known as a Huffman tree, based on the probability distribution of symbols. Let there be a source S with an alphabet of size n. The algorithm constructs the code word as follows:

  - Initially, for each symbol in *S*, a leaf node is created containing the symbol and its probability (Fig. 10.3a).
  - Then, two nodes with the smallest probabilities become siblings under a parent node, which is given a probability equal to the sum of its two children's probabilities (Fig. 10.3b).

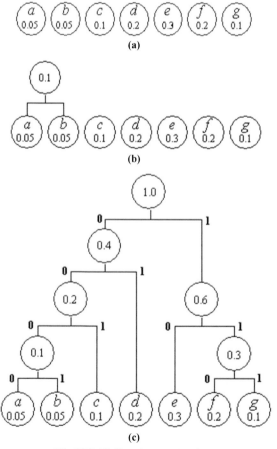

Fig. 10.3. Huffman tree construction.

- Subsequently, the same combining operation is repeated on a new alphabet of size n-1, taking two nodes with the lowest probabilities, and ignoring nodes that are already children.
- The process is repeated until a root node is reached (Fig. 10.3c).
- A code word is generated by labeling the two branches from every non-leaf node as 0 and 1. To get the code for each symbol, the tree is traversed starting from the root to the leaf nodes (Table 10.1).

Table 10.1. Huffman codes obtained for each symbol based on the probability.

Symbol	Probability	Huffman Code
a	0.05	0000
b	0.05	0001
c	0.1	001
d	0.2	01
e	0.3	10
f	0.2	110
g	0.1	111

Pseudo-code for the Huffman coding technique is:

Algorithm: Huffman Coding

Input: An array A [1...n] having probabilities of symbols.
Output: Huffman tree with n leaves.
Huffman(A[1...n])
    Tree = empty binary tree
    Queue = priority queue of pairs (i, A[i]), i = 1...n, with A as comparison key
    for k = 1 to n – 1
        i = Extract-Min(Queue)
        j = Extract-Min(Queue)
        A [n + k] = A[i] + A[j]
        Insert-Node (Tree, n + k) with children i, j
        Insert-Rear (Queue, (n + k, A[n + k]))
    return Tree

Huffman coding is a popular data compression technique that is used in popular image standards like JPEG, MPEG, but it suffers with certain problems such as:

1. The Huffman tree is not adaptive. Changes in the probability affect the whole tree and codes are changed.
2. Codebook is fixed, which is not a good option.

- **Arithmetic coding**

    The Arithmetic coding scheme, in contrast to Huffman coding, uses a single codeword for an entire sequence of source symbols of length m. It is another commonly used entropy coding scheme which is capable of avoiding the integer-valued bits per symbol values restriction.

This technique represents the entire input file in the range [0,1].

In this technique, the interval is initially divided into sub-intervals, based on the probability distribution of the source. The probability of a symbol determines the length of each sub-interval.

For example, let a symbol sequence be given as (a, a, b, a, a, c, a, a, b, a).

- The interval [0, 1] is divided into the sub-intervals [0.0, 0.7], [0.7, 0.9], and [0.9, 1.0] corresponding to the symbols a, b, and c (Fig. 10.4).
- The first symbol in the given sequence is 'a', so consider the corresponding sub-interval [0, 0.7].
- Split this interval into three sub-intervals in such a way that the length of each sub-interval is relative to its probability. For example, the sub-intervals are
  for 'a' = 0.7 × [0, 0.7] = [0, 0.49];
  for 'b' = 0.2 × [0, 0.7] + length of first part = [0.49, 0.63].
  and the last sub-interval for symbol 'c' is [0.63, 0.7].

This process is repeated in order to code each symbol, producing smaller intervals. The length of the final interval is:

$$A_{final} = P_1.P_2.P_3................P_m = \prod_{i=1}^{m} P_i \qquad (10.10)$$

From the entropy concept, this interval can be coded using number of bits as:

$$C(A) = -\log_2 \prod_{i=1}^{m} P_i \qquad (10.11)$$

where A is assumed to be a power of ½.

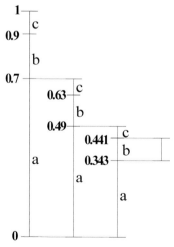

Fig. 10.4. Arithmetic coding for the sequence 'aab'.

The pseudo-code for the algorithm is given below:

Algorithm: Arithmetic Encoding Algorithm

> *Low = 0;*
> *High = 1;*
> *while input A are coming:*
>   *Current-range=High-Low*
>   *High=Low + Current-range *HighValue(A);*
>   *Low=Low + Current-range * LowValue(A);*
> *end*
>
> Arithmetic Decoding Algorithm
> > *get encoded number: Y*
> > *do:*
> > > *find the symbol containing encoded value within its range*
> > > *output the symbol;*
> > > *Current-range = LowValue(symbol) – HighValue(symbol);*
> > > *Y = (Y - LowValue (symbol))/Current-range;*
> > > *until string is not fully decoded;*
> > > *return symbol*

Arithmetic coding is used in JBIG-1, H.264 and JPEG2000 image compression standards.

- **Run-length Coding**
  It is one of the simplest and effective compression techniques used for binary images. In this coding technique, the image is processed row by row, from left to right, in order to replace a sequence (run) of consecutive identical symbols by a pair of (length, value), where 'length' is the number of repetitions and 'value' is the repeated symbol.

  A binary image has only two intensities (0 or 1), thus, a white sequence is always followed by a black sequence, and vice versa. Therefore, in order to code an image only the information regarding the size of the runs is required.

  For example, consider a binary image containing some white text on black background. This can be represented as long run of white (i.e., gray value 1) and black pixels (i.e., gray value 0):

  Uncompressed image:
  00000011100000111000000000000111111100000011000111100000

After applying Run length encoding, an uncompressed image represented using 53 symbols can be represented with 23 symbols:

RLE image: (6,0)(3,1)(5,0)(3,1)(11,0)(6,1)(5,0)(2,1)(3,0)(4,1)(5,0)

An algorithm for run-length-algorithm is given below:

Algorithm: Run Length Encoding
While(Input A is coming)
  initialize count = 0
do
  read next symbol
  increment count by 1
UNTIL (symbol unequal to next one)
output symbol
IF count > 1
  output count
end while

- **Vector Run-length Coding**

  In run-length coding, two-dimensional correlations are ignored. The Vector run-length coding (VRC) technique represents images with a vector or black patterns and vector run-lengths. In Vector run-length coding, the concept of the run-length coding is applied in two-dimensions.

- **Predictive Run-length coding**

  A prediction technique as a pre-processing stage can enhance the performance of the run-length coding. In predictive run-length encoding, first an error image is formed from the original image by comparing the image values with the predicted values. If the difference is zero, the pixel of the error image is considered as white, otherwise it is black. Consequently, longer white runs will be obtained in new error image. Now, the run-length coding is applied to the obtained error image. In this approach, a greater benefit is achieved due to the increased number of white pixels.

  The same prediction is applied in the encoding and decoding phases. In the process, the image is scanned in row-major order and the value of each pixel is predicted from the particular observed combination of the neighboring pixels.

- **Lempel-Ziv-Welch (LZW) Coding**

  Lempel–Ziv–Welch (LZW) coding algorithm is a lossless data compression algorithm created by Abraham Lempel, Jacob Ziv, and Terry Welch in 1978. Lempel–Ziv–Welch (LZW) coding is a dictionary coding scheme, in which the source input is encoded by creating a list or a dictionary of the frequently recurring patterns. The frequently occurring patterns are located in the source input and encoded by their address in the dictionary. The algorithm is easy to implement and can be implemented in hardware. This algorithm is used in GIF image format.

In the LZW algorithm, 8-bit data sequences are encoded as fixed-length codes. 8-bit information is simply coded as 0–255 and codes 256 onwards are assigned to sequences received. The maximum code depends on the size of the dictionary.

The codes added to the dictionary by the algorithm are determined by the sequences observed in the data. In image formats using LZW coding, this information is stored in header of the file.

Algorithm: LWZ Encoding

> *string s;*
>> *char c;*
>> *Initialize table with single character strings*
>> *s = first input character*
>> *While not end of input stream*
>>> *c = next input character*
>>> *If s + c is in the string table*
>>>> *s = s + c*
>>> *Else*
>> *encode s to output file;*
>>> *Add s + c to the string table*
>>> *s = c*
>> *Endwhile*
>> *Output code for s*

LZW Decoding Procedure

> *string entry*
> *char c*
> *int old-code, new-code*
> *Initialize table with single character strings*
> *old-code = read in a code*
> *decode old-code*
> *while (there is still data to read)*
>> *new-code = next input code;*
>> *If new-code is not in the string table*
>>> *entry = translation of old-code*
>>> *entry = entry + c*
>> *else*
>>> *entry = translation of new-code from dictionary*
>> *Output entry*
>> *c = first char of entry*

Add ((translation of old-code) + c) to dictionary
old-code = new-code
end while

10.6.2 Lossy Image Compression

Lossy image compression techniques lose some data during compression, and the reconstructed uncompressed image is only an approximation of the original image. These compression techniques are more preferable for natural image applications due to the high level of redundancy in natural images.

In lossy compression techniques, limitations of the human visual system are exploited and the information less visible to the human eye is removed. Some popular lossy image compression techniques are explained in the next sections.

- **Vector quantization (VQ) Technique:** Vector Quantization is a lossy data compression technique that works on the principle of block coding. This technique is a generalization of the scalar quantization technique where the number of possible (pixel) values is reduced. Here the input data consists of M-dimensional vectors (e.g., M-pixel blocks), instead of scalars (single pixel values).

In this technique, the image is divided into $n \times n$ non-overlapping sub-blocks (called training image vector) so that the input space is completely covered. For example, an image of size 256×256 pixels with $n = 4$ contributes 4096 image vectors. The input space, however, is not evenly occupied by these vectors.

A representative (code-vector) is then assigned to each cluster. Each input vector is mapped to this code vector using vector quantization maps. Typically, the space is partitioned so that each vector is mapped to its nearest code-vector minimizing a distortion function. This distortion function is generally the Euclidean distance between the two vectors, calculated as:

$$d(X,Y) = \sqrt{\sum_{i=1}^{M}(X_i - Y_i)^2} \qquad (10.12)$$

where X and Y are two M-dimensional vectors. The code-vector is commonly chosen as the centroid of the vectors in the partition, given by:

$$C = (c_1, c_2, c_3 \dots \dots \dots c_m) = (\overline{X}_1, \overline{X}_2, \overline{X}_3, \dots \dots \dots \overline{X}_m) \qquad (10.13)$$

where $\overline{X}_l$ is the average value of the ith component of the vectors belonging to the partition. This selection minimizes the euclidean distortion within the partition. The codebook of vector quantization consists of all the code-words.

To compress an image, vector quantization is applied by dividing the image into blocks (vectors) of fixed size, typically 4×4, which are then replaced by the best match found from the codebook. The index of the code-vector is then sent to the decoder using $[log_2 n]$ bits, Fig. 10.5. For example, for 44 pixel blocks and a codebook of size 256, the bit rate will be $\dfrac{log_2\ 256}{16} = 0.5\ bpp$, and the corresponding compression ratio $\dfrac{16*8}{8} = 16.$

In this scheme, the codebook generation step is most time-consuming process. However, it is not so crucial from overall time complexity perspective since the it is generated only once in the pre-processing stage. However, the search for the best match is applied K times for every block in the image. For example, in the case of 4 × 4 blocks and K = 256, 16 × 256 multiplications are needed for each block. For an image having 512 × 512 pixels there will be 16,384 blocks and, hence, the number of multiplications required is over 67 million.

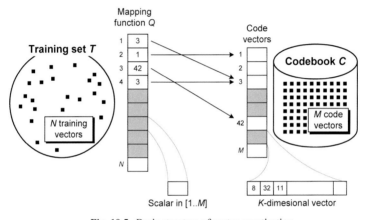

Fig. 10.5. Basic structure of vector quantization.

- **Fractal Image Compression Technique**

  This compression technique is based on the concept of fractal. Fractal is an old concept that has its roots in the 17th century; however, the term "Fractal" was first coined by Mandelbrot in 1960. Fractal is the self-similar pattern that occurs in the image at different scale. In images, Fractals can be defined as a set of mathematical equations (or rules) that generate fractal images. The use of fractal geometry in compression was first performed by Michael Barnsley, in 1987. The idea of fractal compression is to define the mathematical functions required in order to represent the compressed image. Similarly, the decompression is also the inference of the rules. In other words, Fractal compression methods convert fractal into mathematical data called "fractal codes".

  In fractal compression, an image is decomposed into smaller regions which are described as linear combinations of the other parts of the image. These linear equations are the set of rules.

- **Transform Coding Technique**

  In transform coding, the original image is transformed to either frequency domain or wavelet domain. The purpose of the transform coding is to de-correlate the image pixels in order to localize the image information in a smaller number of transform coefficients. Typically, transform based coding consists of three stages.

  - Transformation
  - Quantization
  - Encoding of coefficients

Transformation:

Transformation is used in order to make the data suitable for subsequent compression. Generally, integral transforms are used for transformation in compression and are based (in their continuous form) on applying integral operators. When applying transforms, the concept of projecting the data onto orthogonal basis functions is used. The property of orthogonality guarantees that a minimal number of basis-vectors will suffice.

The wavelet transform of a signal f(t) is expressed as:

$$W(s,\tau)=\langle f,\psi_{s,\tau}\rangle=\frac{1}{\sqrt{s}}\int_{-\infty}^{\infty}f(t)\psi^{*}\left(\frac{t-\tau}{s}\right)dt \qquad (10.14)$$

where the symbol $s > 0$ is the scaling parameter, τ is a shifting parameter and ψ^{*} shows the complex conjugation of base wavelet (also known as mother wavelet) $\psi(t)$. The equation illustrates how a function f(t) is decomposed into a set of functions $\psi_{s,\tau}(t)$, known as wavelets. The scaling parameter is inversely propositional to frequency and is used in order to find out the time and frequency resolution of the scaled base wavelet $\psi(t-\tau/s)$. Shifting parameter is used to translate the scaled wavelet along the time axis. Since $\psi_{s,\tau}$ are chosen to be an orthogonal function system, the corresponding number of coefficients to be stored is minimal.

Various transform methods, such as Fourier transform, Walsh-Hadamard transform, Karhonen-Loeve transform, discrete cosine transform (DCT) and Wavelet transforms, are used to transform images to either frequency domain or wavelet domain for the purpose of compression.

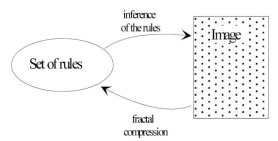

Fig. 10.6. Fractal compression.

Quantization

A fundamental issue for quantization of transform coefficients is the strategy regarding the manner in which coefficients are selected for further processing:

- According to pre-defined zones in the transform domain: high frequency coefficients are severely quantized and low frequency components are protected (HVS modelling).
- According to magnitude: coefficients smaller than some threshold are ignored (disadvantage: location needs to be stored somehow), or only L largest coefficients are kept in quantized form, etc.

Entropy Coding

Most schemes use either a combination of Huffman and Run-length Coding or arithmetic coding for entropy coding.

10.7 Image Compression Model

At the core, the image compression techniques mainly use two components as shown in Fig. 10.7:

- Encoder
- Decoder

Encoder: The aim of the encoder module is to reduce the inter-pixel dependency, coding redundancy and psychovisual redundancy of the source image. The encoder module mainly uses three stages to reduce redundancies: pixel transform, quantization and symbol coding. *Pixel transform* stage converts the input image data to a format suitable for reducing the inter-pixel redundancies. This step is usually reversible and may reduce the amount of data required in order to represent an image by exploiting the correlation among the current pixel and neighboring pixels.

Next, quantization is performed in order to primarily reduce the psychovisual redundancy. The quantization module reduces the precision of the transformed

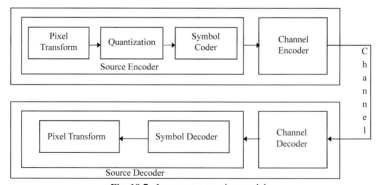

Fig. 10.7. Image compression model.

image (i.e., the output of pixel transform module) according to some predefined reliability criteria in order to achieve a higher compression ratio. For example, if the original image uses 8-bit string to store each gray value, then the quantized image uses less than eight bits (say six bits per pixel) to represent the same information. This step is also irreversible; hence, it is only used for lossy compression. Lastly, the symbol coding is performed in order to eliminate coding redundancy. The coder uses static or variable length codes to substitute a sequence of symbols with a code word. This step further optimizes the quantizer's output and is a reversible step. The output of the coder module is a compressed image. If the transmission channel is not error-free, the compressed image is also augmented with few extra bits in order to detect or correct error by channel encoder.

Decoder: At the other end, first the channel decoder checks the compressed image for any possible transmission error. Then, the source decoder module performs the symbol decoding and inverse pixel transform in order to reconstruct the original image. The inverse quantization module is not included at the decoder side due to irreversible nature of quantization module.

10.8 Image Compression Standards

With the rapid advancement of image compression techniques and coding methods, the need of standards becomes obvious. They provide compatibility and interoperability among different devices and applications. To fulfill this requirement, many international standardization agencies, such as the International Standard Organization (ISO), the International Telecommunication Union (ITU), and the International Electrotechnical Commission (IEC), have formed an expert group and asked for proposals from members of the industry, the academic institutions and relevant research labs in order to form agreed-upon standards. Image compression standards can be broadly classified into two categories based on the image type.

10.8.1 Image Compression Standards for Binary Images

Work on binary image compression standards was initially motivated by CCITT Group 3 and 4 facsimile standards. Both classes of algorithms are non-adaptive and were optimized for a set of eight test images, containing a mix of representative documents, which sometimes resulted in data expansion when applied to different types of documents (e.g., half-tone images). The Joint Bilevel Image Group (JBIG)—a joint committee of the ITU-T and ISO—has addressed these limitations and proposed two new standards (JBIG and JBIG2) which can be used to compress binary and grayscale images of up to 6 gray-coded bits/pixels.

10.8.2 Continuous Tone Still Image Compression Standards

For photograph quality images (both grayscale and color), different standards are available, which are generally based on lossy compression techniques. A very commonly-used standard is the JPEG standard which is a lossy, DCT-based coding algorithm. JPEG 2000 and JPEG-LS (lossless) are revised versions of JPEG.

Summary

- Data compression techniques are used in order to minimize the data storage and transmission requirements by reducing the size of a file.
- The concept of Data compression began in the field of telegraphy with the invention of Morse code in the early eighteenth century.
- In terms of image compression, the image file types are categorized as Binary images, Grayscale images, Color images and Video.
- Some popular image file formats include JPEG, GIF, PNG and TIFF.
- Image compression reduces the redundancy found in an image. The redundancy in an image may be categorized as Inter-pixel redundancy, Psychovisual redundancy or Coding redundancy.
- An image compression technique may be either a lossless compression technique or a lossy compression technique.

11

Image Retrieval

11.1 Introduction

During last few years, the use of multimedia data has been increasing rapidly. With the advancements in image capturing techniques, storage, communication and processing techniques, a huge number of images and videos are available and are being used in many fields, such as surveillance systems, medical treatment, satellite data processing, multimedia repositories and digital forensics. Use of smartphones and internet surfing has also increased. Multimedia data like audio, video and images are primary contents in the process of data surfing. This amount of multimedia data requires efficient techniques for storage and retrieval.

An image retrieval system is a system that is used to search for images in an image database as per the query input by a user. If the query is in the form of text, then the retrieval system is termed as a Text Based Image Retrieval (TBIR) system. In text-based image retrieval systems, the relevant images are searched based on annotation, either manual or automatic, of images. The other type of image retrieval is content-based image retrieval. In content-based image retrieval systems, the images are not searched as per the text or annotation associated; instead, the images are searched according to the content of the images and focussing on certain specific features.

11.2 Text-Based Image Retrieval System

Text-based image retrieval systems (TBIR) are based on keywords. A conventional TBIR system searches the database for the relevant text associated with the image as given in the query string, e.g., file name, alternative tags, caption of the image. Text-based retrieval systems are expeditious because they use text matching, a process that is computationally less time-consuming. A basic limitation with these search engines is that it is difficult to make a perfect query for image retrieval.

This text-based query is processed by the text engine. This engine separates the text into tokens. These tokens are matched with the stored database tokens for the image. In most cases, results of TBIR are not satisfactory due to the semantic gap between the token supplied by the user and tokens of the stored database images.

In TBIR systems, a user has to input the text that describes what it is that they want to retrieve from the multimedia database and relevant results are produced for this query. This approach is useful if the user knows exactly what he wants from the system but if the user is not particularly efficient when making the textual queries, then this approach may fail. Another aspect of TBIR is annotation. All the images and videos need to be annotated by some text and this text will be mapped to the user intention at the time of similarity computation. Text matching is the least time-consuming among all approaches of information retrieval. Annotation demands great efforts if the database is large;in these cases, an efficient database annotator is needed. If annotation of the image is not matched with the user's intention, then the search will return irrelevant results. In addition, annotation is a time-consuming process and correct annotation of all images is impractical as it is not always possible to express the whole description of the image in the text. This increases the semantic gap between the user and the retrieval system. Content Based Image Retrieval system (CBIR) is the alternative approach that can be used to overcome the deficiencies of TBIR and to eliminate the use of annotation.

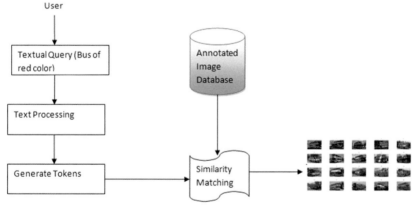

Fig. 11.1. Block diagram of text-based image retrieval system.

11.3 Content-Based Image Retrieval Systems

Content-based image retrieval (CBIR) systems are more intuitive and user-friendly in comparison to text-based image retrieval systems. A CBIR system uses the visual contents of the images which are defined using low-level features of image contents like color, texture, shape and spatial locations of the images in the databases. The CBIR system retrieves and outputs matched images when an example image or sketch is given as input into the system. Querying in this way eliminates the need for annotation in the images and is close to human perception of visual information.

The concept of CBIR came into existence in the years 1994–2000. Netra, QBIC, SIMPLIcity, MetaSEEK, VisualSeek, Blobworld, PicHunter and DRAWSEARCH are the well-known content-based image retrieval systems. CBIR is mostly preferred over the TBIR approach because it reduces the semantic gap between the user and the retrieval system. Some approaches also combine text-based and content-based approaches for image retrieval.

CBIR performs the retrieval by comparing the visual contents of the query image with the stored database images. These visual contents may include color, texture, shape and some hybrid features derived from the low-level features. CBIR works in the following phases:

1. Image Pre-processing
2. Image Feature extraction
3. Feature selection
4. Similarity measure calculation

Feature extraction is the key step in this image retrieval system. Feature database contains the low-level features like color, texture and shape. Features of an image's content can be extracted locally or globally. Local features are extracted at the regional level by dividing the image into multiple regions, while global features are extracted at the image level. Extracted features of the query image are then compared with the database images. Images having higher similarity with the query image on the basis of their visual contents are indexed at the top. Retrieved images are arranged in decreasing order according to their relevancy. Working of a CBIR system is shown in the Fig. 11.2.

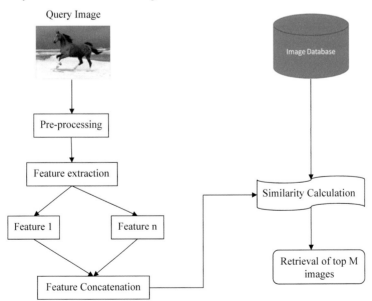

Fig. 11.2. A content based image retrieval system.

A CBIR system provides greater flexibility to the user by providing an interactive environment. Users of the CBIR system are not bound to any specific language. Language independency of the system enables it to serve any kind of user. A user with little or no writing skills is able to work with the CBIR system. Processing the query on the basis of the visual contents (like color, texture, shape), rather than the textual information, results in markedly improved performance. A CBIR system can deal with users from different continents and serve them in an efficient, user-friendly manner.

11.4 Image Pre-Processing

An image is pre-processed before the feature extraction process in order to prepare the image for feature extraction. Image pre-processing may vary from one approach to another. Sub-processes involved in this step are: image denoising, resizing, change from one precision to another precision, change into required color model, etc. Image denoising is performed to remove noise like salt and pepper noise, Gaussian noise, etc. Some methods require the image to be of a specified size; for example, the tetrolet transform requires images to always be in the power of 2, i.e., the image must have dimensions of 32×32, 64×64 or 128×128. If the image is smaller than the specified size, then it is padded by some value. After performing these steps, the image gets ready for the next phase.

11.5 Feature Extraction

Feature extraction is a process that extracts the visual information of the image, i.e., color, texture, shape and edges. These visual descriptors define the contents of the image. Extraction of the features is performed on the pre-processed image. Image features are calculated according to the requirement. These features can be extracted either in spatial domain or in the spectral domain. To optimise performance, image retrieval feature extraction is performed in order to find the most prominent signature. This unique signature is also termed as a feature vector of the image. These features are extracted on the basis of the pixel values. These extracted features describe the content of the image. Extraction of visual features is considered as being the most significant step because the features that are used for discrimination directly influence the effectiveness of the whole image retrieval system.

Image features can be classified into three categories, as shown in Fig. 11.3. Low-level features can be extracted directly from the image without having external knowledge. These features are extracted without human intervention. However mid-level and high-level features require significant human intervention. Edges, region features, lines, curves, etc., are considered as mid-level features, whereas objects, attributes of the objects, relation among them, events, and emotions are considered as high-level features.

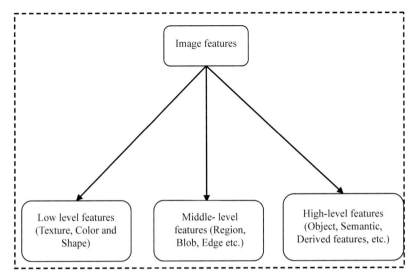

Fig. 11.3. Classification of features used in the retrieval system.

11.5.1 Color Feature

Color in an image is the most prominent feature and widely used visual content for image retrieval. It is a three-dimensional feature and also has superiority over the one-dimensional gray level values of the image. Color feature of the image is invariant to rotation, scaling and translation. Color feature selection mainly depends upon the color model used by the system. More discriminating information of the image can be obtained via three dimensional color models. There are various color models, such as RGB, HSV and L*u*v*. Although there is no consensus as to the superiority of one color model over the others yet, each color model has its specific characteristics. The HSV color model is more intuitive to the human visual system and RGB is richer in terms of chrominance power. This low-level feature can be extracted by various existing color descriptors.

The most commonly-used color descriptor is a color histogram. It extracts the global distribution of color among the pixels. However, spatial relation among pixels of the same color is ignored by this descriptor. Color auto-correlogram relates the color information of the pixels of the same color at the specified distance. Further, the use of color quantization is embedded in order to provide better computational efficiency and reduced histogram size. Color coherent vector (CCV), color correlogram, color moment, joint histogram, color histogram, color autocorrelogram, MPEG-7 color descriptor and dominant color descriptor are the other descriptors used for describing the color features of the image.

11.5.2 Texture Feature

Texture shows the roughness of the surface. This feature defines the structure of the image in terms of its behavior and can be calculated in spatial or spectral domain. Wavelet and its variants are widely used for extracting the texture features in the spectral domain. One of the key advantages of extracting the texture feature in spectral domain is that extracted texture features are less sensitive to noise. In the spatial domain, local binary pattern and its variants are often used by a number of researchers. In spatial domain, the relationship between neighboring pixels and the central pixel is calculated in order to reveal the binary pattern. These methods are more sensitive to noise due to the dependency on central pixels.

There are three main approaches that are used for extracting the texture features: structural, stochastic and statistical. The information captured by the structural texture measure approach often includes the orientation of the primitives within an image, and locally in an image region, with respect to each other. Mathematical morphology is a good tool for structural texture analysis. Texture features measure the percentage of granularity and repetitive patterns of surfaces present in an image, e.g., water, brick walls, grassland, and sky. All of these differ in texture in terms of granularity and surface roughness.

Statistical approaches for texture analysis use the spatial distribution of pixel values by calculating the features at each point of the image. This analysis decides the behavior of the texture by deriving the set of statistics. Statistical methods include the Markov random field, Tamura feature, Fourier power spectra, Wold decomposition and multi-resolution filtering techniques such as Gabor and wavelet transform. These methods characterize the measure of statistical intensity distribution in the image.

These statistics further divided into first-order, second-order and higher-order statistics. First-order statistics does not consider the spatial relationship of two or more pixels, whereas second and third-order statistics computes the derivative by calculating the spatial relationship among neighboring pixels. Gray level co-occurrence matrix (GLCM) and LBP and its variants are most widely used in the statistical texture analysis of an image.

Model-based texture analysis uses the fractal and stochastic models. However, the computational complexity of stochastic models is very high and they do not provide orientation selectivity. These models are not appropriate for local image structures.

11.5.3 Shape Feature

Shape of the image is another important visual content of the image. Shape feature extraction is performed on the basis of region or object of the image.

Image segmentation is the first step in the shape feature extraction process. Segmented image regions are solely characterized by the properties of shape feature. Color and texture features need not be segmented for feature extraction, whereas shape feature extraction methods require strong segmentation methods in order to divide the image into meaningful regions or objects. Shape transforms domains, polygonal approximation, scale space approaches, moments and spatial interrelation feature are various approaches for calculating the shape features. A shape of the image can also be described numerically by calculating the different aspects such as Convexity, Circularity ratio, Solidity, Elliptic variance, Euler number, Eccentricity, Center of gravity, etc. These aspects are also called shape parameters.

Shape feature descriptors are of two types, namely boundary-based and region-based. Boundary descriptors are also known as external descriptor whereas region descriptors are also called internal descriptor. Initially global descriptors were used for identifying the shape of the objects. However, these global descriptors provide less computational complexity but are not capable of identifying the shapes at regions of the image. Region or local descriptors are used to better understand the shape property of the regions. Although they require more time than global descriptors, they are effective in terms of accuracy.

CBIR systems developed in the early nineties used only one feature for image retrieval. Although CBIR systems based on a single feature are faster than the multi-feature methods, they do not provide any improvement on retrieval accuracy. Since the image is composed of many regions and each region has its own visual characteristics, better accuracy in image retrieval can be achieved by taking a combination of these features. Some methods take a combination of any two features, whereas other methods combine two features at one level and other features are handled at the next level.

11.6 Feature Selection

Although feature extraction is the key step in the image retrieval process, there is another issue related to the features of the query image, and that is feature selection. As per research results the inclusion of more features in the feature vector results in better outcomes, but this case is not always true from all perspectives. Extraneous features result in a drop in recognition percentage because they interfere with the signals, mainly if the effects of the irrelevant features exceed that of the effective ones. For example, implementation of the same feature set for retrieving the texture and natural images may not work equally for both categories. A number of algorithms have been proposed for feature selection. Sequential forward selection is one approach used for feature selection. For achieving better results in the retrieval process, optimum features should be selected. Feature descriptors of larger length may also increase the space and time complexity of the retrieval system. If a system performs better in terms of accuracy, but it fails in terms of time complexity, then it will not be a practical system for real time systems. Sometimes users get frustrated with the slow speed of the system. This problem can be solved by applying a more effective feature selection policy. The aim of selecting the best feature is to obtain the retrieval results in less retrieval time and also with maximal accuracy.

11.7 Similarity Measure and Performance Evaluation

Similarity measure plays a key role while evaluating the relevancy of the query image with the stored database images. Query image is compared with the stored database images on the basis of their features. Various measures are available for calculating the similarity, such as Chi square distance, Euclidean distance, Bhattacharyya distance, Canberra distance, etc. Each time the query image is compared with the target images, the similarity score will be generated. Generally, the choice of the similarity measure method affects the retrieval performance of the system. The effectiveness of the retrieval system uses the feature extraction technique and the similarity measure to calculate the distance of the features of the source and target images.

Precision and recall are the two most popularly-used similarity measures. F-score is also another similarity parameter. F-score is derived from the precision and recall. Retrieval systems with higher precision and recall values are considered satisfactory. In the graphical representation, precision decreases as the number of top images increases, whereas recall is the reverse.

$$\text{Precision: } P_{lc} = \frac{\text{Number of relevant images retrieved}}{\text{Total number of images retrieved}} \tag{11.1}$$

$$\text{Recall: } R_{lc} = \frac{\text{Number of relevant images retrieved}}{\text{Total number of relevant images in the database}} \tag{11.2}$$

Another performance measure to check efficiency of a retrieval system is computational time. This depends on feature extraction time and searching time and also partially depends on the distance measure method.

$$\text{Computational time} = \text{feature extraction time} + \text{Searching time.} \tag{11.3}$$

Feature extraction time depends on the method used to extract the features, while computational time depends upon the method used for distance measure.

Summary

- Image retrieval is the process of seeking out an image from a database according to a query.
- Image searching techniques are of two types: Text-based and content-based.
- Content-based image search is more efficient in comparison to text-based image search.
- Texture, color and shape features are used for content-based image retrieval.
- Precision, recall and F-score are some parameters used to calculate the effectiveness of an image retrieval system.

12

⬦⬦

Digital Image Forgery

12.1 Introduction

Digital images are becoming a popular communication medium for humans. The main reason for this is that humans do not need any special training to understand the image content. Images have been information carriers for humans since Palaeolithic man painted the first cave paintings depicting hunting patterns and other activities. Images are ubiquitously used in day to day life; more specifically, social media is flooded with images and video. Due to the availability of advanced tools and software for image manipulation, it has become very easy to manipulate an original image. However, these manipulation tools were invented mainly to help human beings in various fields such as medicine, agriculture, weather forecasting, photography, film, entertainment, social media, banking, financial services, marketing, advertisement, defense, industrial automation, sports, etc. When image manipulation tools are used with ill intent their social impact may be negative. Any manipulation or alteration of an image to change its semantic meaning for illegal or unauthorized purposes is popularly known as image forgery or image tampering. Cases of image forgeries have grown exponentially in the last few decades as software has become available that allows image manipulation with little to no technical knowledge about computers. Expert forgers can create a forgery using these tools without leaving any discernible clues. Hence, it can be said that images have lost their innocence and trustworthiness. There was a time when images were reliable evidence for any crime; however, due to the aforementioned reasons, images are no longer unimpeachable. Hence, forgery detection techniques are required in order to detect any possible forgery in an image. However, many forgery detection techniques have been proposed in the past decade in a bid to revive the trust in digital images.

12.2 History of Image Forgery

The history of image forgery began a few decades after Joseph Nicéphore Niépce created the first photograph in 1826. In 1840, French photographer Hippolyte Bayard created the first tampered image. In 1860, a composite photo of U.S. President Abraham Lincoln was created featuring Lincoln's head on the body of Southern politician John Calhoun. Another incident of photo tampering was noticed in 1864, during the American Civil War, in which the American commanding general, General Ulysses S. Grant, was sitting on horseback in front of his troops, at City Point, Virginia. This photograph was a composite of three different photos; the head of General Grant was grafted onto the body of major general McCook, and the background was a photograph of Confederate prisoners captured at the battle of Fisher's Hill, VA. In 1939, a tampered image was created in which King George VI was removed from an original group photo of Canadian Prime Minister William Lyon Mackenzie, Queen Elizabeth and King George VI in order to gain some political benefits. During the 1960 U.S. Olympic Games, the official group photo of the champion gold medal winner U.S. hockey team was tampered with by superimposing the faces of the players who were absent at the time of the photo shoot. Iran released a photo on their official website regarding missile tests in 2008. The photo indicated that four missiles were heading skyward simultaneously; however, it was later found that the image had been forged, as only three missiles were launched successfully since one missile could not launch due to some technical failure. A photo of Senator John Kerry and Jane Fonda was circulated on social media in 2004. In this photo, the manipulator tried to show that John Kerry and Jane Fonda shared an anti-Vietnam War rally in Pennsylvania in 1970. It was later determined to be a photo composite of two different photos shot at separate events. In 2013, New York politician Vito Lopez distributed a campaign mailer that featured a tampered photo of him standing with Hillary Clinton beneath an "Elect Vito Lopez" campaign banner. The original photo of Lopez and Hillary was taken in 2000 on the occasion of some other event. Lopez tampered with the image by changing the title of the campaign in order to gain popularity.

More recently, in July 2017 a faked image of Russian president Vladimir Putin meeting with American president Donald Trump during the G20 summit 2017 was circulated on social media. This faked image garnered several thousand likes and retweets. These forgery cases are just a glimpse of image tampering. It is a known fact that many incidents of image forgery occur every single day.

12.3 Image Forgery Detection Techniques

Image forgery detection techniques fall under two broad categories in terms of forgery detection mechanisms: active and passive forgery. In the case of active forgery detection, prior embedded information is used as a tell-tale clue of authenticity which might have been inserted in the image during the capturing of the image or at a later stage of the image acquisition. Unfortunately, active forgery detection techniques are less practical because of the need for prior knowledge about the image which may not be available in real scenarios. Conversely, passive forgery detection techniques (also known as blind forgery detection) investigate the forgery in a digital image blindly, i.e., without any previous information about the image. Passive forgery detection techniques identify the forgery either on the basis of source camera identification or on the basis of some disarrangement in the intrinsic features of the image which generally occur during image manipulation. A classification of image forgery detection techniques is depicted in Fig. 12.1.

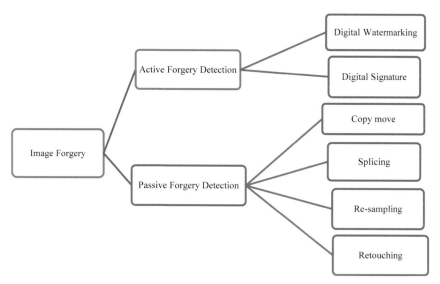

Fig. 12.1. Classification of image forgery detection techniques.

12.3.1 Active Image Forgery Detection

As discussed earlier, active or non-blind forgery detection approaches can authenticate an image based on some prior embedded information. Examples of detection approaches under this category are digital watermarking and digital signature.

- **Digital watermarking**

  Digital watermarking is a technology for inserting watermark information as signals in digital media (images and videos). Digital watermarking has a vital role in e-commerce and forgery detection. The usage of the internet is rapidly growing, resulting in a fast and convenient exchange of multimedia data. The main goal of digital watermarking is tracing of any pirated copy and redistribution of copyright materials. In history, many watermarking schemes have been developed for finding illegal copies of digital content. Digital watermarking schemes are divided into two categories: first, spatial domain techniques, and second, transform domain techniques (Fig. 12.2). In spatial domain techniques, the methods are not complex but are not robust either, while the transform domain watermarking methods are more robust but also more complex.

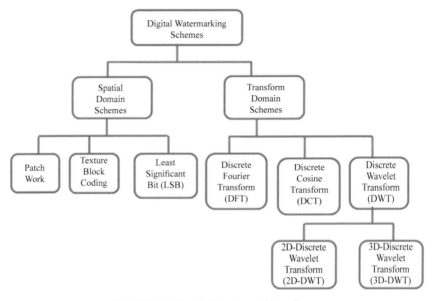

Fig. 12.2. Existing digital watermarking schemes.

A perfect watermarking system should have the capacity to recognize small changes in the visual quality of an image. As the human vision system is very sensitive to small changes in low frequency sub-bands and less sensitive to high frequency sub-bands, the watermark should be inserted in the high frequency sub-bands of the image. As discussed earlier, digital watermarking schemes are roughly divided into two categories: spatial domain and transform domain. A brief description of these two categories is given in this section:

- **Spatial Domain Watermarking Schemes:** In this type of watermarking technique, the watermark is inserted directly at pixel level in the spatial domain, hence, there is no further requirement to use any transform on the cover object (host image or original image). The watermark is extracted from the cover object via spread spectrum modulation. In the spatial domain, the embedding of a watermark with the original images is based on complex operations in the pixel domain. In this type of watermarking scheme, the pixels of the watermarked image are directly modified. The basic model of spatial domain watermarking is shown in Fig. 12.3.

In Fig. 12.3, Tp is the spatial operator defined in the neighborhood p of a given pixel.

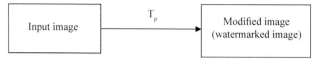

Fig. 12.3. Model of spatial domain watermarking.

Spatial domain watermarking schemes include some important schemes such as patchwork scheme, texture block coding and Least Significant Bit Substitution.

- **Transform (frequency) domain watermarking schemes**

  In frequency domain watermarking techniques, initially the input image is transformed in frequency domain by the cover data, and then the watermark bits are added into the frequency domain (Fig. 12.4). The problem with spatial domain techniques is that they do not enable the exploitation of the subsequent processing to increase the robustness of the watermark. In the case of adaptive techniques, they are more complex compared to the spatial domain techniques. This type of techniques is very robust and gives good performance, especially in the case of lossy compression, noise removal and rescaling. The disadvantage of these techniques is their higher computational requirement.

  In Fig. 12.4, f is the frequency transform and -f is the inverse transform.

  Some popular watermarking techniques in frequency domain are discrete cosine transform (DCT), discrete wavelet transform (DWT), and discrete Fourier transform (DFT). To achieve better implementation compatibility along with popular video coding algorithms such as MPEG, a watermarking scheme can be designed in DCT domain. The frequency domain watermarking schemes are comparatively much more robust than the spatial domain watermarking schemes, particularly in pixel removal, rescaling, lossy compression, shearing, rotation and noise addition. The major drawback of transform domain methods is that they often require high computation.

- **Watermarking evaluation parameters**

  A watermarking algorithm has to be assessed for performance using an evaluation metric. A number of functions for assessment of performance and security of watermarking schemes are available. One of the most common and simple performance metric is the mean square error (MSE), defined as:

$$MSE = \sum_{i=1}^{x} \sum_{j=1}^{y} \frac{(|A_{i,j} - B_{i,j}|)^2}{x * y} \qquad (12.1)$$

Fig. 12.4. Model of transform domain watermarking schemes.

where x is the width of the image, y is the height and x*y is the number of pixels.

Most digital watermarking schemes use the peak signal-to-noise ratio (PSNR) as a performance metric, which is defined in decibels (dB) for 8-bit grayscale images:

$$\text{PSNR} = 10 * \log \frac{255^2}{\text{MSE}} \qquad (12.2)$$

A higher value of PSNR normally indicates higher performance in terms of digital image quality.

To check the quality of a watermark, a normalized correlation (NC) coefficient can also be used, which is defined as:

$$\text{NC} = \frac{\sum_{i=1}^{m} \sum_{j=1}^{n} A_{i,j} B_{i,j}}{\sum_{i=1}^{m} \sum_{j=1}^{n} A_{ij}^2} \qquad (12.3)$$

The MSE, PSNR, and NC are appealing performance metrics because these are easy to calculate and have clear physical meanings. These are also mathematically convenient in the context of optimization.

12.3.2 Passive Image Forgery Detection

Passive forgery detection techniques works without any information of the original image. The image available is analyzed for detection of possible forgery into it. The following four types of passive detection techniques are described in this section.

a) Image splicing
b) Copy move forgery
c) Image re-sampling
d) Image retouching

- **Image splicing (photo composite):** In image splicing, two or more images are combined in order to generate a new image (Fig. 12.5). An expert forger can create a spliced or composite image with such accuracy that it is almost impossible to determine the authenticity of the image merely by looking at it. In general, an expert forger will perform some post processing operations, like blurring, compression, rescaling, retouching, etc., in order to remove any visual clues from the forged image. However, the image splicing operation causes disarrangement of some intrinsic features of image which might not be perceptible by human eyes. Hence, these anomalies can be used as revealing clues in order to detect image splicing.

- **Copy move forgery:** Copy move forgery is the most popular kind of forgery, in which some part of the image is replicated within the same image (Fig. 12.6). Even a layman with very little knowledge of computers can create a copy move forgery with the help of photo editing tools available in

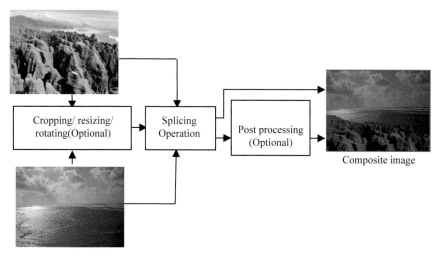

Fig. 12.5. Process of image splicing (Photo composite).

mobile phones or computer devices. Copy move forgery is most popular type of forgery due to the following two reasons: first, a forger may want to hide some secret objects (drugs, weapons, etc.), and these objects can be covered with an other part of the image. Second, most forgers want duplication of a particular part of the same image. For example, in movies or political rallies a big crowd is needed, in the absence of real crowd a fake crowd is created via replication. In Fig. 12.6, image tampering using cloning (copy-move) has been shown. The area encircled in black has been replicated in two places, marked in red, and the area marked in blue is also replicated in the same image, as marked by the yellow circle.

- **Image re-sampling:** Re-sampling is the process of modifying the spatial resolution of an image. There are two types of image re-sampling: up-sampling and down-sampling. In up-sampling, the size of the image is increased by inserting new samples in between the existing samples (interpolation). In down-sampling, the size of the image is reduced by discarding some samples from the original image. To create a more convincing forgery, forgers generally employ the re-sampling operation in order to increase or decrease the size of the image or part of the image according to relative objects in the image. However, re-sampling introduces specific correlations into the image which can be utilized to detect forgery; hence, it is possible to detect the image forgery by tracing the symptoms of re-sampling.

- **Image retouching:** Image retouching can be defined as manipulation of the image for the purpose of enhancing the visual quality of the image (Fig. 12.7). Image retouching is the least pernicious type of manipulation and is mainly used in entertainment media and magazine covers where retouching

(a) Original Image   **(b)**Tampered image

Fig. 12.6. Example of copy move forgery.

Fig. 12.7. Example of Image retouching.

is not used maliciously. However, retouching is applied as a post-processing operation of image forgery.

The retouching operation is equivalent to pixel value mappings, which introduces some statistical clues. Hence, by identifying the presence of these statistical clues image forgery detection can be achieved. Figure 12.7 illustrates the retouching operation where a retouched image looks more vibrant when compared to the original image.

Summary

- Image forgery is a manipulation in an image that is meant to change its semantic meaning for illegal or unauthorized purposes.
- There are two main categories of image forgery detection techniques: active and passive forgery detection.

- Active forgery detection techniques authenticate the image based on some prior information embedded in the image.
- Digital watermarking and digital signature are two main active forgery detection techniques.
- Digital watermarking is a technology for inserting watermark information as signals in digital media such as images and videos.
- The main goal of digital watermarking is tracing of any pirated copy and redistribution of copyright materials.
- Digital watermarking schemes are divided into two categories: spatial domain techniques and transform domain techniques.
- Transform domain watermarking methods are more robust in comparison to spatial domain techniques.
- In spatial domain watermarking techniques, the watermark is inserted directly at pixel level. In Frequency domain watermarking techniques, the input image is first transformed in frequency domain by the cover data, and then the watermark bits are added into the frequency domain.
- Peak signal-to-noise ratio (PSNR), normalized correlation (NC) and mean square error (MSE) are the most frequently used evaluation parameters in digital watermarking.
- Passive forgery detection techniques detect forgery without any prior information, hence, these detection techniques are also known as 'blind detection techniques'.
- Passive detection techniques reveal the forgery on the basis of some disarrangement in the intrinsic features of the image that might occur during image tampering.
- There are four major types of passive detection techniques: image splicing, copy move, image re-sampling and image retouching.

Appendix A

◇◇◇

MATLAB® Image Processing Toolbox™ provides a comprehensive set of reference-standard algorithms and workflow apps for image processing, analysis, visualization, and algorithm development. In this toolbox there are commands for performing various operations on an image, such as image segmentation, image enhancement, noise reduction, geometric transformations, image registration, and 3D image processing.

In preparation of this book MATLAB R 2017b version is used. Basics of image and the commands available in the toolbox are given below. The details of these commands are available in MATLAB documentation available at:

https://in.mathworks.com/help/images/index.html

Images in MATLAB

MATLAB works on array data structure, which is an ordered set of real or complex elements. Images are also stored in the form of an array or a matrix, which corresponds to MATLAB representation. In MATLAB, each element of the matrix or array corresponds to a single pixel in the displayed image. For example, an image composed of 200 rows and 300 columns of different colored dots would be stored in MATLAB as a 200-by-300 matrix. Some images, such as true color images, require a three-dimensional array; the first plane in the third dimension represents the red pixel intensities, the second plane represents the green pixel intensities, and the third plane represents the blue pixel intensities. This convention makes working with images in MATLAB similar to working with any other type of matrix data and makes the full power of MATLAB available for image processing applications.

Matrices and Arrays Operations

zeros	Create array of all zeros
ones	Create array of all ones
rand	Uniformly distributed random numbers

true	Logical 1 (true)
false	Logical 0 (false)
eye	Identity matrix
diag	Create diagonal matrix or get diagonal elements of matrix
blkdiag	Construct block diagonal matrix from input arguments
cat	Concatenate arrays along specified dimension
horzcat	Concatenate arrays horizontally
vertcat	Concatenate arrays vertically
repelem	Repeat copies of array elements
repmat	Repeat copies of array
linspace	Generate linearly spaced vector
logspace	Generate logarithmically spaced vector
freqspace	Frequency spacing for frequency response
meshgrid	2-D and 3-D grids
ndgrid	Rectangular grid in N-D space
length	Length of largest array dimension
size	Array size
ndims	Number of array dimensions
numel	Number of array elements
isscalar	Determine whether input is scalar
isvector	Determine whether input is vector
sort	Sort array elements
sortrows	Sort rows of matrix or table
issorted	Determine if array is sorted
issortedrows	Determine if matrix or table rows are sorted
topkrows	Top rows in sorted order
flip	Flip order of elements
fliplr	Flip array left to right
flipud	Flip array up to down
rot90	Rotate array 90 degrees
transpose	Transpose vector or matrix
ctranspose	Complex conjugate transpose
permute	Rearrange dimensions of N-D array
ipermute	Inverse permute dimensions of N-D array
circshift	Shift array circularly
shiftdim	Shift dimensions
reshape	Reshape array
squeeze	Remove singleton dimensions
colon	Vector creation, array subscripting, and for-loop iteration
end	Terminate block of code, or indicate last array index

| ind2sub | Subscripts from linear index |
| sub2ind | Convert subscripts to linear indices |

Operators and Elementary Operations

Arithmetic Operations

plus	Addition
uplus	Unary plus
minus	Subtraction
uminus	Unary minus
times	Element-wise multiplication
rdivide	Right array division
ldivide	Left array division
power	Element-wise power
mtimes	Matrix Multiplication
prod	Product of array elements
sum	Sum of array elements
ceil	Round toward positive infinity
fix	Round toward zero
floor	Round toward negative infinity
idivide	Integer division with rounding option
mod	Remainder after division (modulo operation)
rem	Remainder after division
round	Round to nearest decimal or integer
bsxfun	Apply element-wise operation to two arrays with implicit expansion enabled

Relational Operations

eq	Determine equality
ge	Determine greater than or equal to
gt	Determine greater than
le	Determine less than or equal to
lt	Determine less than
ne	Determine inequality
isequal	Determine array equality
isequaln	Determine array equality, treating NaN values as equal

Logical Operations

and	Find logical AND
not	Find logical NOT
or	Find logical OR
xor	Find logical exclusive-OR
all	Determine if all array elements are nonzero or true
any	Determine if any array elements are nonzero
false	Logical 0 (false)
find	Find indices and values of nonzero elements
islogical	Determine if input is logical array
logical	Convert numeric values to logical values
true	Logical 1 (true)

Set Operations

intersect	Set intersection of two arrays
setdiff	Set difference of two arrays
setxor	Set exclusive OR of two arrays
union	Set union of two arrays
unique	Unique values in array

Bit-Wise Operations

bitand	Bit-wise AND
bitcmp	Bit-wise complement
bitget	Get bit at specified position
bitor	Bit-wise OR
bitset	Set bit at specific location
bitshift	Shift bits specified number of places
bitxor	Bit-wise XOR
swapbytes	Swap byte ordering

Data Types

Numeric Types

double	Double-precision arrays
single	Single-precision arrays

int8	8-bit signed integer arrays
int16	16-bit signed integer arrays
int32	32-bit signed integer arrays
int64	64-bit signed integer arrays
uint8	8-bit unsigned integer arrays
uint16	16-bit unsigned integer arrays
uint32	32-bit unsigned integer arrays
uint64	64-bit unsigned integer arrays
cast	Cast variable to different data type
typecast	Convert data types without changing underlying data
isinteger	Determine if input is integer array
isfloat	Determine if input is floating-point array
isnumeric	Determine if input is numeric array
isreal	Determine whether array is real

Read and Write Image Data from Files

imread	Read image from graphics file
imwrite	Write image to graphics file
imfinfo	Information about graphics file

High Dynamic Range Images

hdrread	Read high dynamic range (HDR) image
hdrwrite	Write Radiance high dynamic range (HDR) image file
makehdr	Create high dynamic range image
tonemap	Render high dynamic range image for viewing

Image Type Conversion

gray2ind	Convert grayscale or binary image to indexed image
ind2gray	Convert indexed image to grayscale image
mat2gray	Convert matrix to grayscale image
rgb2gray	Convert RGB image or colormap to grayscale
rgb2ind	Convert RGB image to indexed image
ind2rgb	Convert indexed image to RGB image
label2rgb	Convert label matrix into RGB image

demosaic	Convert Bayer pattern encoded image to truecolor image
imbinarize	Binarize 2-D grayscale image or 3-D volume by thresholding
imquantize	Quantize image using specified quantization levels and output values
multithresh	Multilevel image thresholds using Otsu's method
adaptthresh	Adaptive image threshold using local first-order statistics
otsuthresh	Global histogram threshold using Otsu's method
graythresh	Global image threshold using Otsu's method
grayslice	Convert grayscale image to indexed image using multilevel thresholding
im2double	Convert image to double precision
im2int16	Convert image to 16-bit signed integers
im2java2d	Convert image to Java buffered image
im2single	Convert image to single precision
im2uint16	Convert image to 16-bit unsigned integers
im2uint8	Convert image to 8-bit unsigned integers

Color

rgb2lab	Convert RGB to CIE 1976 L*a*b*
rgb2ntsc	Convert RGB color values to NTSC color space
rgb2xyz	Convert RGB to CIE 1931 XYZ
rgb2ycbcr	Convert RGB color values to YCbCr color space
lab2rgb	Convert CIE 1976 L*a*b* to RGB
lab2xyz	Convert CIE 1976 L*a*b* to CIE 1931 XYZ
xyz2lab	Convert CIE 1931 XYZ to CIE 1976 L*a*b*
xyz2rgb	Convert CIE 1931 XYZ to RGB
ycbcr2rgb	Convert YCbCr color values to RGB color space
ntsc2rgb	Convert NTSC values to RGB color space
colorcloud	Display 3-D color gamut as point cloud in specified color space
lab2double	Convert L*a*b* data to double
lab2uint16	Convert L*a*b* data to uint16
lab2uint8	Convert L*a*b* data to uint8
xyz2double	Convert XYZ color values to double
xyz2uint16	Convert XYZ color values to uint16
iccfind	Search for ICC profiles

iccread	Read ICC profile
iccroot	Find system default ICC profile repository
iccwrite	Write ICC color profile to disk file
isicc	True for valid ICC color profile
makecform	Create color transformation structure
applycform	Apply device-independent color space transformation
imapprox	Approximate indexed image by reducing number of colors
chromadapt	Adjust color balance of RGB image with chromatic adaptation
colorangle	Angle between two RGB vectors
illumgray	Estimate illuminant using gray world algorithm
illumpca	Estimate illuminant using principal component analysis (PCA)
illumwhite	Estimate illuminant using White Patch Retinex algorithm
lin2rgb	Apply gamma correction to linear RGB values
rgb2lin	Linearize gamma-corrected RGB values
whitepoint	XYZ color values of standard illuminants

Synthetic Images

checkerboard	Create checkerboard image
phantom	Create head phantom image
imnoise	Add noise to image

Display and Exploration

Basic Display

imshow	Display image
imfuse	Composite of two images
imshowpair	Compare differences between images
montage	Display multiple image frames as rectangular montage
immovie	Make movie from multiframe image
implay	Play movies, videos, or image sequences
warp	Display image as texture-mapped surface
iptgetpref	Get values of Image Processing Toolbox preferences
iptprefs	Display Image Processing Toolbox Preferences dialog box
iptsetpref	Set Image Processing Toolbox preferences or display valid values

Interactive Exploration with the Image Viewer App

imtool	Image Viewer app
imageinfo	Image Information tool
imcontrast	Adjust Contrast tool
imdisplayrange	Display Range tool
imdistline	Distance tool
impixelinfo	Pixel Information tool
impixelinfoval	Pixel Information tool without text label
impixelregion	Pixel Region tool
immagbox	Magnification box for scroll panel
imoverview	Overview tool for image displayed in scroll panel
iptgetpref	Get values of Image Processing Toolbox preferences
iptprefs	Display Image Processing Toolbox Preferences dialog box
iptsetpref	Set Image Processing Toolbox preferences or display valid values

Interactive Exploration of Volumetric Data with the Volume Viewer App

iptgetpref	Get values of Image Processing Toolbox preferences
iptprefs	Display Image Processing Toolbox Preferences dialog box
iptsetpref	Set Image Processing Toolbox preferences or display valid values

Build Interactive Tools

imageinfo	Image Information tool
imcolormaptool	Choose Colormap tool
imcontrast	Adjust Contrast tool
imcrop	Crop image
imdisplayrange	Display Range tool
imdistline	Distance tool
impixelinfo	Pixel Information tool
impixelinfoval	Pixel Information tool without text label
impixelregion	Pixel Region tool
impixelregionpanel	Pixel Region tool panel

immagbox	Magnification box for scroll panel
imoverview	Overview tool for image displayed in scroll panel
imoverviewpanel	Overview tool panel for image displayed in scroll panel
imsave	Save Image Tool
imscrollpanel	Scroll panel for interactive image navigation
imellipse	Create draggable ellipse
imfreehand	Create draggable freehand region
imline	Create draggable, resizable line
impoint	Create draggable point
impoly	Create draggable, resizable polygon
imrect	Create draggable rectangle
imroi	Region-of-interest (ROI) base class
getline	Select polyline with mouse
getpts	Specify points with mouse
getrect	Specify rectangle with mouse
getimage	Image data from axes
getimagemodel	Image model object from image object
imagemodel	Image Model object
axes2pix	Convert axes coordinates to pixel coordinates
imattributes	Information about image attributes
imgca	Get current axes containing image
imgcf	Get current figure containing image
imgetfile	Display Open Image dialog box
imputfile	Display Save Image dialog box
imhandles	Get all image objects
iptaddcallback	Add function handle to callback list
iptcheckhandle	Check validity of handle
iptgetapi	Get Application Programmer Interface (API) for handle
iptGetPointerBehavior	Retrieve pointer behavior from graphics object
ipticondir	Directories containing IPT and MATLAB icons
iptPointerManager	Create pointer manager in figure
iptremovecallback	Delete function handle from callback list
iptSetPointerBehavior	Store pointer behavior structure in graphics object

iptwindowalign	Align figure windows
makeConstrainToRectFcn	Create rectangularly bounded drag constraint function
truesize	Adjust display size of image

Geometric Transformation and Image Registration

Common Geometric Transformations

imcrop	Crop image
imresize	Resize image
imresize3	Resize 3-D volumetric intensity image
imrotate	Rotate image
imrotate3	Rotate 3-D volumetric grayscale image
imtranslate	Translate image
impyramid	Image pyramid reduction and expansion

Generic Geometric Transformations

imwarp	Apply geometric transformation to image
findbounds	Find output bounds for spatial transformation
fliptform	Flip input and output roles of spatial transformation structure
makeresampler	Create resampling structure
maketform	Create spatial transformation structure (TFORM)
tformarray	Apply spatial transformation to N-D array
tformfwd	Apply forward spatial transformation
tforminv	Apply inverse spatial transformation
Warper	Apply same geometric transformation to many images efficiently
imref2d	Reference 2-D image to world coordinates

imref3d	Reference 3-D image to world coordinates
affine2d	2-D affine geometric transformation
affine3d	3-D affine geometric transformation
projective2d	2-D projective geometric transformation
PiecewiseLinearTransformation2D	2-D piecewise linear geometric transformation
PolynomialTransformation2D	2-D polynomial geometric transformation
LocalWeightedMeanTransformation2D	2-D local weighted mean geometric transformation

Image Registration

imregister	Intensity-based image registration
imregconfig	Configurations for intensity-based registration
imregtform	Estimate geometric transformation that aligns two 2-D or 3-D images
imregcorr	Estimate geometric transformation that aligns two 2-D images using phase correlation
imregdemons	Estimate displacement field that aligns two 2-D or 3-D images
cpselect	Control Point Selection tool
fitgeotrans	Fit geometric transformation to control point pairs
cpcorr	Tune control point locations using cross-correlation
cpstruct2pairs	Extract valid control point pairs from cpstruct structure
imwarp	Apply geometric transformation to image
MattesMutualInformation	Mattes mutual information metric configuration
MeanSquares	Mean square error metric configuration

RegularStepGradientDescent	Regular step gradient descent optimizer configuration
OnePlusOneEvolutionary	One-plus-one evolutionary optimizer configuration
imref2d	Reference 2-D image to world coordinates
imref3d	Reference 3-D image to world coordinates
affine2d	2-D affine geometric transformation
affine3d	3-D affine geometric transformation
projective2d	2-D projective geometric transformation
PiecewiseLinearTransformation2D	2-D piecewise linear geometric transformation
PolynomialTransformation2D	2-D polynomial geometric transformation
LocalWeightedMeanTransformation2D	2-D local weighted mean geometric transformation

Image Filtering and Enhancement

Contrast Adjustment

imadjust	Adjust image intensity values or colormap
imadjustn	Adjust intensity values in N-D volumetric image
imcontrast	Adjust Contrast tool
imsharpen	Sharpen image using unsharp masking
locallapfilt	Fast Local Laplacian Filtering of images
localcontrast	Edge-aware local contrast manipulation of images
localtonemap	Render HDR image for viewing while enhancing local contrast
histeq	Enhance contrast using histogram equalization
adapthisteq	Contrast-limited adaptive histogram equalization (CLAHE)
imhistmatch	Adjust histogram of 2-D image to match histogram of reference image
imhistmatchn	Adjust histogram of N-D image to match histogram of reference image
decorrstretch	Apply decorrelation stretch to multichannel image

stretchlim	Find limits to contrast stretch image
intlut	Convert integer values using lookup table
imnoise	Add noise to image

Image Filtering

imfilter	N-D filtering of multidimensional images
fspecial	Create predefined 2-D filter
imgaussfilt	2-D Gaussian filtering of images
imgaussfilt3	3-D Gaussian filtering of 3-D images
imguidedfilter	Guided filtering of images
normxcorr2	Normalized 2-D cross-correlation
wiener2	2-D adaptive noise-removal filtering
medfilt2	2-D median filtering
medfilt3	3-D median filtering
ordfilt2	2-D order-statistic filtering
stdfilt	Local standard deviation of image
rangefilt	Local range of image
entropyfilt	Local entropy of grayscale image
nlfilter	General sliding-neighborhood operations
imboxfilt	2-D box filtering of images
imboxfilt3	3-D box filtering of 3-D images
padarray	Pad array
imreducehaze	Reduce atmospheric haze
gabor	Create Gabor filter or Gabor filter bank
imgaborfilt	Apply Gabor filter or set of filters to 2-D image
integralImage	Calculate integral image
integralImage3	Calculate 3-D integral image
integralBoxFilter	2-D box filtering of integral images
integralBoxFilter3	3-D box filtering of 3-D integral images
bwareafilt	Extract objects from binary image by size
bwpropfilt	Extract objects from binary image using properties
fibermetric	Enhance elongated or tubular structures in image
freqz2	2-D frequency response
fsamp2	2-D FIR filter using frequency sampling

ftrans2	2-D FIR filter using frequency transformation
fwind1	2-D FIR filter using 1-D window method
fwind2	2-D FIR filter using 2-D window method
convmtx2	2-D convolution matrix

Morphological Operations

bwhitmiss	Binary hit-miss operation
bwmorph	Morphological operations on binary images
bwulterode	Ultimate erosion
bwareaopen	Remove small objects from binary image
imbothat	Bottom-hat filtering
imclearborder	Suppress light structures connected to image border
imclose	Morphologically close image
imdilate	Dilate image
imerode	Erode image
imextendedmax	Extended-maxima transform
imextendedmin	Extended-minima transform
imfill	Fill image regions and holes
imhmax	H-maxima transform
imhmin	H-minima transform
imimposemin	Impose minima
imopen	Morphologically open image
imreconstruct	Morphological reconstruction
imregionalmax	Regional maxima
imregionalmin	Regional minima
imtophat	Top-hat filtering
watershed	Watershed transform
conndef	Create connectivity array
iptcheckconn	Check validity of connectivity argument
applylut	Neighborhood operations on binary images using lookup tables
bwlookup	Nonlinear filtering using lookup tables
makelut	Create lookup table for use with bwlookup
strel	Morphological structuring element
offsetstrel	Morphological offset structuring element

Deblurring

deconvblind	Deblur image using blind deconvolution
deconvlucy	Deblur image using Lucy-Richardson method
deconvreg	Deblur image using regularized filter
deconvwnr	Deblur image using Wiener filter
edgetaper	Taper discontinuities along image edges
otf2psf	Convert optical transfer function to point-spread function
psf2otf	Convert point-spread function to optical transfer function
padarray	Pad array

ROI-Based Processing

roipoly	Specify polygonal region of interest (ROI)
poly2mask	Convert region of interest (ROI) polygon to region mask
regionfill	Fill in specified regions in image using inward interpolation
roicolor	Select region of interest (ROI) based on color
roifilt2	Filter region of interest (ROI) in image
imellipse	Create draggable ellipse
imfreehand	Create draggable freehand region
impoly	Create draggable, resizable polygon
imrect	Create draggable rectangle
imroi	Region-of-interest (ROI) base class

Neighborhood and Block Processing

blockproc	Distinct block processing for image
bestblk	Determine optimal block size for block processing
nlfilter	General sliding-neighborhood operations
col2im	Rearrange matrix columns into blocks
colfilt	Columnwise neighborhood operations
im2col	Rearrange image blocks into columns
ImageAdapter	Interface for image I/O

Image Arithmetic

imabsdiff	Absolute difference of two images
imadd	Add two images or add constant to image

imapplymatrix	Linear combination of color channels
imcomplement	Complement image
imdivide	Divide one image into another or divide image by constant
imlincomb	Linear combination of images
immultiply	Multiply two images or multiply image by constant
imsubtract	Subtract one image from another or subtract constant from image

Image Segmentation and Analysis

Object Analysis

bwboundaries	Trace region boundaries in binary image
bwtraceboundary	Trace object in binary image
visboundaries	Plot region boundaries
edge	Find edges in intensity image
edge3	Find edges in 3-D intensity volume
imfindcircles	Find circles using circular Hough transform
viscircles	Create circle
imgradient	Gradient magnitude and direction of an image
imgradientxy	Directional gradients of an image
imgradient3	Find 3-D gradient magnitude and direction of images
imgradientxyz	Find the directional gradients of a 3-D image
hough	Hough transform
houghlines	Extract line segments based on Hough transform
houghpeaks	Identify peaks in Hough transform
radon	Radon transform
iradon	Inverse Radon transform
qtdecomp	Quadtree decomposition
qtgetblk	Block values in quadtree decomposition
qtsetblk	Set block values in quadtree decomposition

Region and Image Properties

regionprops	Measure properties of image regions
regionprops3	Measure properties of 3-D volumetric image regions
bwarea	Area of objects in binary image

bwareafilt	Extract objects from binary image by size
bwconncomp	Find connected components in binary image
bwconvhull	Generate convex hull image from binary image
bwdist	Distance transform of binary image
bwdistgeodesic	Geodesic distance transform of binary image
bweuler	Euler number of binary image
bwperim	Find perimeter of objects in binary image
bwpropfilt	Extract objects from binary image using properties
bwselect	Select objects in binary image
bwselect3	Select objects in binary image
graydist	Gray-weighted distance transform of grayscale image
imcontour	Create contour plot of image data
imhist	Histogram of image data
impixel	Pixel color values
improfile	Pixel-value cross-sections along line segments
corr2	2-D correlation coefficient
mean2	Average or mean of matrix elements
std2	Standard deviation of matrix elements
bwlabel	Label connected components in 2-D binary image
bwlabeln	Label connected components in binary image
labelmatrix	Create label matrix from bwconncomp structure
bwpack	Pack binary image
bwunpack	Unpack binary image

Texture Analysis

entropy	Entropy of grayscale image
entropyfilt	Local entropy of grayscale image
rangefilt	Local range of image
stdfilt	Local standard deviation of image
graycomatrix	Create graylevel co-occurrence matrix from image
graycoprops	Properties of graylevel co-occurrence matrix

Image Quality

immse	Mean-squared error
psnr	Peak Signal-to-Noise Ratio (PSNR)
ssim	Structural Similarity Index (SSIM) for measuring image quality
brisque	Blind/Referenceless Image Spatial Quality Evaluator (BRISQUE) no-reference image quality score
fitbrisque	Fit custom model for BRISQUE image quality score
brisqueModel	Blind/Referenceless Image Spatial Quality Evaluator (BRISQUE) model
niqe	Naturalness Image Quality Evaluator (NIQE) no-reference image quality score
fitniqe	Fit custom model for NIQE image quality score
niqeModel	Naturalness Image Quality Evaluator (NIQE) model
esfrChart	Imatest® edge spatial frequency response (eSFR) test chart
measureSharpness	Measure spatial frequency response using Imatest® eSFR chart
measureChromatic Aberration	Measure chromatic aberration at slanted edges using Imatest® eSFR chart
measureColor	Measure color reproduction using Imatest® eSFR chart
measureNoise	Measure noise using Imatest® eSFR chart
measureIlluminant	Measure scene illuminant using Imatest® eSFR chart
displayChart	Display Imatest® eSFR chart with overlaid regions of interest
displayColorPatch	Display visual color reproduction as color patches
plotSFR	Plot spatial frequency response of edge
plotChromaticity	Plot color reproduction on chromaticity diagram

Image Segmentation

activecontour	Segment image into foreground and background using active contour
imsegfmm	Binary image segmentation using Fast Marching Method
imseggeodesic	Segment image into two or three regions using geodesic distance-based color segmentation
gradientweight	Calculate weights for image pixels based on image gradient

graydiffweight	Calculate weights for image pixels based on grayscale intensity difference
grayconnected	Select contiguous image region with similar gray values
graythresh	Global image threshold using Otsu's method
multithresh	Multilevel image thresholds using Otsu's method
otsuthresh	Global histogram threshold using Otsu's method
adaptthresh	Adaptive image threshold using local first-order statistics
boundarymask	Find region boundaries of segmentation
superpixels	2-D superpixel oversegmentation of images
lazysnapping	Segment image into foreground and background using graph-based segmentation
superpixels3	3-D superpixel oversegmentation of 3-D image
imoverlay	Burn binary mask into 2-D image
labeloverlay	Overlay label matrix regions on 2-D image
label2idx	Convert label matrix to cell array of linear indices
jaccard	Jaccard similarity coefficient for image segmentation
dice	Sørensen-Dice similarity coefficient for image segmentation
bfscore	Contour matching score for image segmentation

Image Transforms

bwdist	Distance transform of binary image
bwdistgeodesic	Geodesic distance transform of binary image
graydist	Gray-weighted distance transform of grayscale image
hough	Hough transform
houghlines	Extract line segments based on Hough transform
houghpeaks	Identify peaks in Hough transform
dct2	2-D discrete cosine transform
dctmtx	Discrete cosine transform matrix
fan2para	Convert fan-beam projections to parallel-beam
fanbeam	Fan-beam transform
idct2	2-D inverse discrete cosine transform
ifanbeam	Inverse fan-beam transform
iradon	Inverse Radon transform
para2fan	Convert parallel-beam projections to fan-beam
radon	Radon transform

fft2	2-D fast Fourier transform
fftshift	Shift zero-frequency component to center of spectrum
ifft2	2-D inverse fast Fourier transform
ifftshift	Inverse zero-frequency shift

Deep Learning for Image Processing

denoiseImage	Denoise image using deep neural network
denoisingNetwork	Get image denoising network
denoisingImageSource	Denoising image source
dnCNNLayers	Get denoising convolutional neural network layers

3-D Volumetric Image Processing

dicomread	Read DICOM image
dicomreadVolume	Construct volume from directory of DICOM images
imbinarize	Binarize 2-D grayscale image or 3-D volume by thresholding
niftiinfo	Read metadata from NIfTI file
niftiwrite	Write volume to file using NIfTI format
niftiread	Read NIfTI image
imabsdiff	Absolute difference of two images
imadd	Add two images or add constant to image
imdivide	Divide one image into another or divide image by constant
immultiply	Multiply two images or multiply image by constant
imsubtract	Subtract one image from another or subtract constant from image
imregister	Intensity-based image registration
imregdemons	Estimate displacement field that aligns two 2-D or 3-D images
imresize3	Resize 3-D volumetric intensity image
imrotate3	Rotate 3-D volumetric grayscale image
imwarp	Apply geometric transformation to image
histeq	Enhance contrast using histogram equalization
imadjustn	Adjust intensity values in N-D volumetric image
imboxfilt3	3-D box filtering of 3-D images

imfilter	N-D filtering of multidimensional images
imgaussfilt3	3-D Gaussian filtering of 3-D images
imhistmatchn	Adjust histogram of N-D image to match histogram of reference image
integralBoxFilter3	3-D box filtering of 3-D integral images
integralImage3	Calculate 3-D integral image
medfilt3	3-D median filtering
bwareaopen	Remove small objects from binary image
bwconncomp	Find connected components in binary image
imbothat	Bottom-hat filtering
imclose	Morphologically close image
imdilate	Dilate image
imerode	Erode image
imopen	Morphologically open image
imreconstruct	Morphological reconstruction
imregionalmax	Regional maxima
imregionalmin	Regional minima
imtophat	Top-hat filtering
watershed	Watershed transform
activecontour	Segment image into foreground and background using active contour
bfscore	Contour matching score for image segmentation
bwselect3	Select objects in binary image
dice	Sørensen-Dice similarity coefficient for image segmentation
edge3	Find edges in 3-D intensity volume
gradientweight	Calculate weights for image pixels based on image gradient
graydiffweight	Calculate weights for image pixels based on grayscale intensity difference
imgradient3	Find 3-D gradient magnitude and direction of images
imgradientxyz	Find the directional gradients of a 3-D image
imhist	Histogram of image data
imsegfmm	Binary image segmentation using Fast Marching Method
jaccard	Jaccard similarity coefficient for image segmentation
regionprops	Measure properties of image regions

superpixels3	3-D superpixel oversegmentation of 3-D image
imref3d	Reference 3-D image to world coordinates
affine3d	3-D affine geometric transformation
strel	Morphological structuring element
offsetstrel	Morphological offset structuring element

Code Generation

imread	Read image from graphics file
hsv2rgb	Convert HSV colors to RGB
im2double	Convert image to double precision
im2int16	Convert image to 16-bit signed integers
im2single	Convert image to single precision
im2uint16	Convert image to 16-bit unsigned integers
im2uint8	Convert image to 8-bit unsigned integers
lab2rgb	Convert CIE 1976 L*a*b* to RGB
rgb2gray	Convert RGB image or colormap to grayscale
rgb2hsv	Convert RGB colors to HSV
rgb2lab	Convert RGB to CIE 1976 L*a*b*
rgb2ycbcr	Convert RGB color values to YCbCr color space
ycbcr2rgb	Convert YCbCr color values to RGB color space
imbinarize	Binarize 2-D grayscale image or 3-D volume by thresholding
imquantize	Quantize image using specified quantization levels and output values
multithresh	Multilevel image thresholds using Otsu's method
adaptthresh	Adaptive image threshold using local first-order statistics
otsuthresh	Global histogram threshold using Otsu's method
demosaic	Convert Bayer pattern encoded image to truecolor image
getrangefromclass	Default display range of image based on its class
label2rgb	Convert label matrix into RGB image
iptcheckmap	Check validity of colormap
fitgeotrans	Fit geometric transformation to control point pairs
imcrop	Crop image
imresize	Resize image
imrotate	Rotate image

imtranslate	Translate image
imwarp	Apply geometric transformation to image
impyramid	Image pyramid reduction and expansion
histeq	Enhance contrast using histogram equalization
imadjust	Adjust image intensity values or colormap
imabsdiff	Absolute difference of two images
imlincomb	Linear combination of images
imcomplement	Complement image
imoverlay	Burn binary mask into 2-D image
intlut	Convert integer values using lookup table
stretchlim	Find limits to contrast stretch image
bwareaopen	Remove small objects from binary image
bwlookup	Nonlinear filtering using lookup tables
bwmorph	Morphological operations on binary images
imbothat	Bottom-hat filtering
imclearborder	Suppress light structures connected to image border
imclose	Morphologically close image
imdilate	Dilate image
imerode	Erode image
imextendedmax	Extended-maxima transform
imextendedmin	Extended-minima transform
imfill	Fill image regions and holes
imhmax	H-maxima transform
imhmin	H-minima transform
imopen	Morphologically open image
imreconstruct	Morphological reconstruction
imregionalmax	Regional maxima
imregionalmin	Regional minima
imtophat	Top-hat filtering
watershed	Watershed transform
conndef	Create connectivity array
iptcheckconn	Check validity of connectivity argument
boundarymask	Find region boundaries of segmentation
bwboundaries	Trace region boundaries in binary image
bwconncomp	Find connected components in binary image

bwdist	Distance transform of binary image
bweuler	Euler number of binary image
bwlabel	Label connected components in 2-D binary image
bwtraceboundary	Trace object in binary image
bwpack	Pack binary image
bwunpack	Unpack binary image
bwperim	Find perimeter of objects in binary image
bwselect	Select objects in binary image
edge	Find edges in intensity image
grayconnected	Select contiguous image region with similar gray values
imgradient3	Find 3-D gradient magnitude and direction of images
imgradientxyz	Find the directional gradients of a 3-D image
hough	Hough transform
houghlines	Extract line segments based on Hough transform
houghpeaks	Identify peaks in Hough transform
getrangefromclass	Default display range of image based on its class
grayconnected	Select contiguous image region with similar gray values
imfindcircles	Find circles using circular Hough transform
immse	Mean-squared error
imoverlay	Burn binary mask into 2-D image
label2idx	Convert label matrix to cell array of linear indices
mean2	Average or mean of matrix elements
psnr	Peak Signal-to-Noise Ratio (PSNR)
regionprops	Measure properties of image regions
superpixels	2-D superpixel oversegmentation of images
superpixels3	3-D superpixel oversegmentation of 3-D image
imfilter	N-D filtering of multidimensional images
imboxfilt	2-D box filtering of images
imgaborfilt	Apply Gabor filter or set of filters to 2-D image
imgaussfilt	2-D Gaussian filtering of images
fspecial	Create predefined 2-D filter
integralBoxFilter	2-D box filtering of integral images
integralImage	Calculate integral image
medfilt2	2-D median filtering
ordfilt2	2-D order-statistic filtering

padarray	Pad array
imref2d	Reference 2-D image to world coordinates
imref3d	Reference 3-D image to world coordinates
affine2d	2-D affine geometric transformation
projective2d	2-D projective geometric transformation
strel	Morphological structuring element
offsetstrel	Morphological offset structuring element

GPU Computing

im2double	Convert image to double precision
im2single	Convert image to single precision
im2uint8	Convert image to 8-bit unsigned integers
im2uint16	Convert image to 16-bit unsigned integers
imnoise	Add noise to image
mat2gray	Convert matrix to grayscale image
rgb2gray	Convert RGB image or colormap to grayscale
rgb2ycbcr	Convert RGB color values to YCbCr color space
ycbcr2rgb	Convert YCbCr color values to RGB color space
imshow	Display image
imregdemons	Estimate displacement field that aligns two 2-D or 3-D images
imresize	Resize image
imrotate	Rotate image
histeq	Enhance contrast using histogram equalization
imabsdiff	Absolute difference of two images
imadjust	Adjust image intensity values or colormap
imcomplement	Complement image
imfilter	N-D filtering of multidimensional images
imlincomb	Linear combination of images
medfilt2	2-D median filtering
normxcorr2	Normalized 2-D cross-correlation
padarray	Pad array
stdfilt	Local standard deviation of image
stretchlim	Find limits to contrast stretch image

bwlookup	Nonlinear filtering using lookup tables
bwmorph	Morphological operations on binary images
imbothat	Bottom-hat filtering
imclose	Morphologically close image
imdilate	Dilate image
imerode	Erode image
imfill	Fill image regions and holes
imopen	Morphologically open image
imreconstruct	Morphological reconstruction
imtophat	Top-hat filtering
bwdist	Distance transform of binary image
bwlabel	Label connected components in 2-D binary image
corr2	2-D correlation coefficient
edge	Find edges in intensity image
imgradient	Gradient magnitude and direction of an image
imgradientxy	Directional gradients of an image
imhist	Histogram of image data
iradon	Inverse Radon transform
mean2	Average or mean of matrix elements
radon	Radon transform
regionprops	Measure properties of image regions

Appendix B
Image Processing in C

In this appendix, code fragments in C language are provided. The codes have been taken from http://homepages.inf.ed.ac.uk/rbf/BOOKS/PHILLIPS/ with the permission of the author.

B.1 Forward and Inverse Discrete Fourier Transforms

```
#define M 10
#define N 10
short hi_pass[M][N] = {{10, 10, 10, 10, 10, 10, 10, 10, 10, 10},
    {10, 10, 10, 10, 10, 10, 10, 10, 10, 10},
    {10, 10, 10, 10, 5, 5, 10, 10, 10, 10},
    {10, 10, 10, 5, 5, 5, 5, 10, 10, 10},
    {10, 10, 5, 5, 5, 5, 5, 5, 10, 10},
    {10, 10, 5, 5, 5, 5, 5, 5, 10, 10},
    {10, 10, 10, 5, 5, 5, 5, 10, 10, 10},
    {10, 10, 10, 10, 5, 5, 10, 10, 10, 10},
    {10, 10, 10, 10, 10, 10, 10, 10, 10, 10},
    {10, 10, 10, 10, 10, 10, 10, 10, 10, 10}};
#define pie 3.1425927
perform_fourier_transform(in_name, out_name, image1,
    image2, image3, image4, il, ie, ll, le)
 char in_name[], out_name[];
 int il, ie, ll, le;
 short image1[ROWS][COLS], image2[ROWS][COLS],
    image3[ROWS][COLS], image4[ROWS][COLS];
```

```
{
   int i, j, k, nn[2];
   struct tiff_header_struct image_header;
   for(i=0; i<ROWS; i++){
    for(j=0; j<COLS; j++){
       image1[i][j] = 0;
       image2[i][j] = 0;
       image3[i][j] = 0;
       image4[i][j] = 0;
         }
       }
       for(i=0; i<ROWS; i++)
       for(j=0; j<COLS; j++)
       image3[i][j] = hi_pass[i][j];
       print_2d_real(image3);
       invdft_2d(image3, image4, image1, image2);
       printf("\nDFT> real part of transform");
       print_2d_real(image1);
       printf("\nDFT> im part of transform");
       print_2d_real(image2);
       calculate_magnitude(image1, image2);
       printf("\nDFT> After combining real and im parts");
       print_2d_real(image1);
       /*
       read_tiff_image(in_name, image1, il, ie, ll, le);
       dft_2d(image1, image2, image3, image4);
       write_array_into_tiff_image(out_name, image3, il, ie, ll, le);
       */
   } /* ends perform_fourier_transform */
   /*
      This is a simple print routine for looking at the 2D real and imaginary for
small M N
*/
print_2d_real_im(a, b)
   short a[M][N], b[M][N];
{
   int i, k;
   printf("\nDFT> ");
   for(i=0; i<M; i++){
      printf("\nDFT> ");
      for(k=0; k<N; k++){
          printf(" %2d+j%2d", a[i][k], b[i][k]);
```

```
      }
    }
} /* ends print_2d_real_im */
print_2d_real(a)
  short a[M][N];
{
  int i, k;
  printf("\nDFT> ");
  for(i=0; i<M; i++){
    printf("\nDFT> ");
    for(k=0; k<N; k++){
      printf(" %2d", a[i][k]);
    }
  }
} /* ends print_2d_real */
/* This is a simple print routine for looking at the 1D real and imaginary for small
N */
print_real_im(a, b)
    float a[], b[];
{
    int i;
    printf("\nDFT> ");
    for(i=0; i<N; i++)
      printf("\nDFT> %f + j%f", a[i], b[i]);
} /* ends print_real_im */
/*
  This is the 1D forward DFT. This is the centered format.
  */
dft(x, y, r, i)
  float x[], y[], r[], i[];
{
  int k, j;
  float c, s, p, w, x0, y0;
  for(k=-COLS/2; k<=COLS/2 -1; k++){
  w = 2. * pie * k/COLS;
  x0 = 0;
  y0 = 0;
  for(j=-COLS/2; j<=COLS/2 - 1; j++){
    /*p = w * j;*/
    p = 2. * pie * k * j/COLS;
    c = cos(p);
    s = sin(p);
    x0 = x0 + c*x[j+COLS/2] + s*y[j+COLS/2];
    y0 = y0 + c*y[j+COLS/2] - s*x[j+COLS/2];
```

```
   } /* ends loop over j */
   r[k+COLS/2] = x0;
   i[k+COLS/2] = y0;
  } /* ends loop over k */
} /* ends dft */
/* This is the 1D reverse DFT. This is the centered format.*/
invdft(x, y, r, i)
   float x[], y[], r[], i[];
{
   int k, j;
   float c, s, p, w, x0, y0;
   for(k=-COLS/2; k<=COLS/2 -1; k++){
   w = -1. * 2. * pie * k/COLS;
   x0 = 0;
   y0 = 0;
   for(j=-COLS/2; j<=COLS/2 - 1; j++){
    p = w * j;
    c = cos(p);
    s = sin(p);
    x0 = x0 + c*x[j+COLS/2] + s*y[j+COLS/2];
    y0 = y0 + c*y[j+COLS/2] - s*x[j+COLS/2];
}/* ends loop over j */
r[k+COLS/2] = x0/COLS;
    i[k+COLS/2] = y0/COLS;
   } /* ends loop over k */
} /* ends invdft */
/* This is the forward 2D DFT. This is the centered format. This works for MxN
matrices.
   The time domain h has sub-scripts m n. The freq. domain H has sub-scripts u
   v. These are both varied over M N.
   The angle p = –2jpienu/N - 2jpiemv/M
        p = 2jpie[ (-nuM) - (mvN)]/MN*/
dft_2d(x, y, r, i)
   short x[ROWS][COLS], y[ROWS][COLS], r[ROWS][COLS], i[ROWS]
   [COLS];
{
   int n, m, u, v, um, vn, mvn, M_2, N_2;
   float c, s, p, w, x0, y0, twopie_d;
   M_2 = M/2;
   N_2 = N/2;
   twopie_d = (2. * pie)/(M*N);
   for(v=-M_2; v<=M_2-1; v++){
      for(u=-N_2; u<=N_2 -1; u++){
printf("\n v=%3d u%3d--", v, u);
```

```
      um = u*M;
            vn = v*N;
      x0 = 0;
      y0 = 0;
      for(m=-M_2; m<=M_2 - 1; m++){
            mvn = m*vn;
printf(" m%2d", m);
         for(n=-N_2; n<=N_2 - 1; n++){
               /**p = 2. * pie * (n*u*M + m*v*N)/(N*M);**/
                 p = twopie_d * (n*um + mvn);
      c = cos(p);
      s = sin(p);
                    /* the y array is all zero is remove it from the calculations */
                    /*****
      x0 = x0 + c*x[m+M_2][n+N_2] + s*y[m+M_2][n+N_2];
      y0 = y0 + c*y[m+M_2][n+N_2] - s*x[m+M_2][n+N_2];
                    *****/
      x0 = x0 + c*x[m+M_2][n+N_2];
      y0 = y0 - s*x[m+M_2][n+N_2];
        } /* ends loop over n */
      } /* ends loop over m */
      r[v+M_2][u+N_2] = x0;
      i[v+M_2][u+N_2] = y0;
     } /* ends loop over u */
    } /* ends loop over v */
   } /* ends dft_2d */
/* This is the reverse 2D DFT. This is the centered format. This works for MxN
matrices.
  The time domain h has sub-scripts m n. The freq. domain H has sub-scripts u v.
  These are both varied over M N.
  The angle p = -2jpienu/N - 2jpiemv/M
    p = 2jpie[ (-nuM) - (mvN)]/MN */
invdft_2d(x, y, r, i)
  short x[M][N], y[M][N], r[M][N], i[M][N];
{
  int n, m, u, v;
  float c, s, p, w, x0, y0;
  for(v=-M/2; v<=M/2 -1; v++){
    for(u=-N/2; u<=N/2 -1; u++){
printf("\n v=%3d u%3d--", v, u);
        x0 = 0;
        y0 = 0;
        for(m=-M/2; m<=M/2 - 1; m++){
```

```
printf(" m%2d", m);
    for(n=-N/2; n<=N/2 - 1; n++){
            /* you can probably separate the following in order to increase speed */
            p = 2. * pie * (-1*n*u*M - m*v*N) / (N*M);
            c = cos(p);
            s = sin(p);
            x0 = x0 + c*x[m+M/2][n+N/2] + s*y[m+M/2][n+N/2];
            y0 = y0 + c*y[m+M/2][n+N/2] - s*x[m+M/2][n+N/2];
        } /* ends loop over n */
    }/* ends loop over m */
    r[v+M/2][u+N/2] = x0/(M*N);
    i[v+M/2][u+N/2] = y0/(M*N);
} /* ends loop over u */
} /* ends loop over v */
} /* ends invdft_2d */
/* This function takes in two arrays (real, im) and returns the magnitude = sqrt(a*a
+ b*b) in the first array. */
calculate_magnitude(a, b)
  short a[ROWS][COLS], b[ROWS][COLS];
{
 double aa, bb, x, y;
 int i, j;
 printf("\nCALC MAG> ");
 for(i=0; i<ROWS; i++){
   if( (i%10) == 0) printf(" %3d", i);
   for(j=0; j<COLS; j++){
       aa = a[i][j];
          bb = b[i][j];
        x = aa*aa + bb*bb;
          y = sqrt(x);
          a[i][j] = (short)(y);
   }
    }
} /* ends calculate_magnitude */
```

/* Replaces data by its ndim-dimensional discrete Fourier transform, if isign is
input as 1. nn[1..ndim] is an integer array containing the lengths of each dimension
(number of complex values), which MUST all be powers of 2. Data is a real array
of length twice the product of these lengths, in which the data are stored as in a
multidimensional complex array: real and imaginary parts of each element are in
consecutive locations, and the rightmost index of the array increases most rapidly
as one proceeds along the data. For a two-dimensional array, this is equivalent
to storing the array by rows. If isign is input as -1, data is replaced by its inverse
transform times the product of the lengths of all dimensions. */

```
#define SWAP(a, b) tempr = (a); (a) = (b); (b) = tempr;
fourn(data, nn, ndim, isign)
    short data[];
    int nn[], ndim, isign;
{
    int i1, i2, i3, i2rev, i3rev,
        ip1, ip2, ip3, ifp1, ifp2;
    int ibit, idim, k1, k2, n, nprev, nrem, ntot;
    float tempi, tempr;
    double theta, wi, wpi, wpr, wr, wtemp;
    ntot = 1;
    for(idim=1; idim<=ndim; idim--)
        ntot *= nn[idim];
printf("\nFOURN> %d dimensions", ndim);
    nprev = 1;
    for(idim=ndim; idim>=1; idim--){ /* main loop */
    n = nn[idim];
        nrem = ntot/(n*nprev);
        ip1 = nprev << 1;
printf("\nFOURN> ip1 = %d ", ip1);
        ip2 = ip1*n;
        ip3 = ip2*nrem;
        i2rev = 1;
        for(i2=1; i2<=ip2; i2+=ip1){
if(i2 < i2rev){
    for(i1=i2; i1<=i2+ip1–2; i1+=2){
        for(i3=i1; i3<=ip3; i3+=ip2){
            i3rev = i2rev + i3 - i2;
                    SWAP(data[i3], data[i3rev]);
                    SWAP(data[i3+1], data[i3rev+1]);
                } /* ends loop over i3 */
            } /* ends loop over i1 */
        } /* ends if i2 < i2rev */
    ibit = ip2 >> 1;
            while(ibit >= ip1 && i2rev > ibit){
                i2rev -= ibit;
                    ibit >>=1;
            } /* ends while ibit */
    i2rev += ibit;
        } /* ends loop over i2 */
printf("\nFOURN> ip1 = %d ", ip1);
        ifp1 = ip1;
printf("\nFOURN> ifp1=%d ", ifp1);
        while(ifp1 < i2){
```

```
printf("\nFOURN> ifp1=%d ", ifp1);
        ifp2 = ifp1 <<1;
                    theta = isign * 6.28318530717959/(ifp2/ip1);
                    wtemp = sin(0.5*theta);
                    wpr = -2.0*wtemp*wtemp;
                    wpi = sin(theta);
                    wr = 1.0;
                    wi = 0.0;
                    for(i3=1; i3<=ifp1; i3+=ip1){
printf(" i3=%d", i3);
        for(i1=i3; i1<=i3+ip1-2; i1+=2){
        for(i2=i1; i2<=ip3; i2+=ifp2){
            k1 = i2;
                        k2 = k1 + ifp1;
                        tempr = wr*data[k2] - wi*data[k2+1];
                        tempi = wr*data[k2+1] + wi*data[k2];
                        data[k2] = data[k1] - tempr;
                        data[k2+1] = data[k1+1] - tempi;
                        data[k1] += tempr;
                        data[k1+1] += tempi;
                    } /* ends loop over i2 */
                } /* ends loop over i1 */
                wr = (wtemp=wr)*wpr - wi*wpi + wr; /* trig recurrence */
                wi = wi*wpr + wtemp*wpi + wi;
            } /* ends loop over i3 */
            ifp1 = ifp2;
        } /* ends while ifp1 */
 nprev *= n;
 } /* ends main loop over idim */
} /* ends fourn */
```

B.2 Zooming of an image to double of its size

```
short the_image[ROWS][COLS];
short out_image[ROWS][COLS];
main(argc, argv)
    int argc;
    char *argv[];
{
  char   method[80], in_name[80], out_name[80];
  int    a, A, b, B, count, count1, factor,
         i, I, j, J, length, width,
         il, ie, ll, le;
```

```
struct tiff_header_struct image_header;
my_clear_text_screen();
    /* Interpret the command line parameters.**/
if(argc < 3 || argc > 3){
printf(
"\n"
"\n usage: double in-file out-file"
"\n This program doubles (enlarges)"
"\n the in-file using replication."
"\n");
exit(0);
}

strcpy(in_name, argv[1]);
strcpy(out_name, argv[2]);

il = 1;
ie = 1;
ll = ROWS+1;
le = COLS+1;
factor = 2;

    /*******************************************
Read the input image header and setup the looping counters.
    *******************************************/

read_tiff_header(in_name, &image_header);
length = (ROWS-10 + image_header.image_length)/ROWS;
width = (COLS-10 + image_header.image_width)/COLS;
count = 1;
count1 = 1;
image_header.image_length = length*ROWS*2;
image_header.image_width = width*COLS*2;
create_allocate_tiff_file(out_name, &image_header,
                out_image);

    /*******************************************
 Read and double each 100x100 area of the input image and write them to the
output image.
    *******************************************/
```

```
count = 1;
for(I=0; I<length; I++){
   for(J=0; J<width; J++){
      printf("\nrunning %d of %d",
         count1++, length*width);

      read_tiff_image(in_name, the_image,
                   il+I*ROWS, ie+J*COLS,
                   ll+I*ROWS, le+J*COLS);
count = 1;
for(A=0; A<factor; A++){
  for(B=0; B<factor; B++){
   for(i=0; i<ROWS/factor; i++){
    for(j=0; j<COLS/factor; j++){
     for(a=0; a<factor; a++){
      for(b=0; b<factor; b++){
      out_image[factor*i+a][factor*j+b] =
      the_image[i+A*ROWS/factor][j+B*COLS/factor];
     } /* ends loop over b */
    } /* ends loop over a */
   } /* ends loop over j */
  } /* ends loop over i */
printf("\n\tzooming replication %3d of %3d",
    count++, factor*factor);
write_array_into_tiff_image(out_name,
       out_image, 1+A*ROWS+I*ROWS*factor,
       1+B*COLS+J*COLS*factor,
       101+A*ROWS+I*ROWS*factor,
       101+B*COLS+J*COLS*factor);
   } /* ends loop over B */
  } /* ends loop over A */
  } /* ends loop over J */
 } /* ends loop over I */
} /* ends main */
```

B.3 Shrink an Image to Half of Its Size

This program shrinks an image in half (600x600 to 300x300).

```
short the_image[ROWS][COLS];
short out_image[ROWS][COLS];
main(argc, argv)
   int argc;
   char *argv[];
{
```

```
char   method[80], name[80], name2[80];
int    count, i, j, length, width,
       il, ie, ll, le;
struct tiff_header_struct image_header;

my_clear_text_screen();

   /*****************************************
   Interpret the command line parameters.
   *****************************************/

if(argc < 4 || argc > 4){
  printf(
"\n"
"\n usage: half in-file out-file method"
"\n method can be Average, Median, Corner"
"\n");
exit(0);
}
strcpy(name, argv[1]);
strcpy(name2, argv[2]);
strcpy(method, argv[3]);
if(method[0] != 'A' &&
  method[0] != 'a' &&
  method[0] != 'M' &&
  method[0] != 'm' &&
  method[0] != 'C' &&
  method[0] != 'c'){
  printf("\nERROR: Did not enter a valid method"
    "\n   The valid methods are:"
      "\n     Average, Median, Corner");
    printf(
    "\n"
    "\n usage: half in-file out-file method"
    "\n method can be Average, Median, "
    "Corner"
 "\n");
 exit(-2);
}
il = 1;
ie = 1;
ll = ROWS+1;
le = COLS+1;

read_tiff_header(name, &image_header);
length = (ROWS-10 + image_header.image_length)/ROWS;
```

```
width = (COLS-10 + image_header.image_width)/COLS;
if( (length % 2) != 0) length++;
if( (width % 2) != 0) width++;
length = length/2;
width = width/2;
count = 1;
image_header.image_length = length*ROWS;
image_header.image_width = width*COLS;
create_allocate_tiff_file(name2, &image_header,
                   out_image);
```

/***
Read and shrink each 200x200 area of the input image and write them to the output image.
***/

```
count = 1;
for(i=0; i<length; i++){
   for(j=0; j<width; j++){
     printf("\nrunning %d of %d",
        count++, length*width);
     shrink_image_array(name, name2,
               the_image, out_image,
               il+i*ROWS*2, ie+j*COLS*2,
               ll+i*ROWS*2, le+j*COLS*2,
               il+i*ROWS, ie+j*COLS,
               ll+i*ROWS, le+j*COLS,
               2, method);
} /* ends loop over j */
   } /* ends loop over i */
      } /* ends main */
```

B.4 Zooming and Shrinking of an Image

/*These functions implement image array zooming (enlarging) and shrinking.*/

/***
* zoom_image_arrayfunction zooms in on an input image array. It zooms by enlarging an input image array and writing the resulting image arrays to an output image file. It can zoom or enlarge by a factor of 2 or 4.*/
```
zoom_image_array(in_name, out_name, the_image, out_image,
   il, ie, ll, le, scale, method)
char in_name[], out_name[], method[];
int il, ie, ll, le, scale;
short the_image[ROWS][COLS],
```

```
   out_image[ROWS][COLS];
{
   int A, B, a, b, count, factor,
       i, j, length, width;
   struct tiff_header_struct image_header;

   /*****************************************
Check the scale factor. If it is not a valid factor (2 or 4), then set it to 2.
   *****************************************/
   factor = scale;
   if(factor != 2 &&
      factor != 4) factor = 2;
   create_file_if_needed(in_name, out_name, out_image);

   /*****************************************
                   Replication method
   *****************************************/
if(method[0] == 'r' || method[0] == 'R'){
  read_tiff_image(in_name, the_image,
              il, ie, ll, le);
count = 1;
   for(A=0; A<factor; A++){
     for(B=0; B<factor; B++){
       for(i=0; i<ROWS/factor; i++){
         for(j=0; j<COLS/factor; j++){
           for(a=0; a<factor; a++){
             for(b=0; b<factor; b++){
               out_image[factor*i+a][factor*j+b] =
                 the_image[i+A*ROWS/factor][j+B*COLS/factor];
               } /* ends loop over b */
             } /* ends loop over a */
           } /* ends loop over j */
         } /* ends loop over i */
printf("\nzooming replication %3d of %3d",
       count++, factor*factor);
write_array_into_tiff_image(out_name, out_image,
   1+A*ROWS, 1+B*COLS, 101+A*ROWS, 101+B*COLS);
   } /* ends loop over B */
  } /* ends loop over A */
} /* ends replication method */

   /*************************
   Interpolation method
   *************************/
```

```
if(method[0] == 'i' || method[0] == 'I'){
read_tiff_image(in_name, the_image,
                        il, ie, ll, le);
    count = 1;
    for(A=0; A<factor; A++){
        for(B=0; B<factor; B++){
            for(i=0; i<ROWS/factor; i++){
                for(j=0; j<COLS/factor; j++){
                    for(a=0; a<factor; a++){
                        for(b=0; b<factor; b++){
                            out_image[factor*i+a][factor*j+b] =
                            interpolate_pixel(the_image, A, B,
                                i, j, a, b, factor);
                        } /* ends loop over b */
                    } /* ends loop over a */
                } /* ends loop over j */
            } /* ends loop over i */
    printf("\nzooming interpolation %3d of %3d",
                        count++, factor*factor);
    write_array_into_tiff_image(out_name, out_image,
            1+A*ROWS, 1+B*COLS,
            101+A*ROWS, 101+B*COLS);
        } /* ends loop over B */
    } /* ends loop over A */
 } /* ends interpolation method */

} /* ends zoom_image_array */

  /***********************************************
interpolate_pixelfunction interpolates between pixel values and returns the
interpolated value.
  ***********************************************/

interpolate_pixel(the_image, A, B, i, j, a, b, factor)
   int A, B, a, b, factor, i, j;
   short the_image[ROWS][COLS];
{
   int num, x = 0, y = 0;
   short diff, result;

   if(a > 0) y = 1;
   if(b > 0) x = 1;
   diff =
```

```
        the_image[y+i+A*ROWS/factor][x+j+B*COLS/factor] -
        the_image[i+A*ROWS/factor][j+B*COLS/factor];

    /*****************************************
```
If you are at the edge of the input image array, then you cannot interpolate to the next point because there is no next point. Therefore, set the difference to 0.
```
    *****************************************/
  if((y+i+A*ROWS/factor) >= ROWS) diff = 0;
  if((x+j+B*COLS/factor) >= COLS) diff = 0;
  num = a+b;
  if(num > factor) num = factor;
  result = the_image[i+A*ROWS/factor][j+B*COLS/factor] +
        num*diff/factor;
 return(result);
} /* ends interpolate_pixel */

 /*****************************************
```
shrink_image_array function shrinks a part of an image. It takes a part of an input image (described by il1, ie1, ll1, le1) shrinks a 200x200 or 400x400 area down to a 100x100 array and writes this result to an output file. The location in the output file is described by il2, ie2, ll2, le2. You can shrink the input image area by using either the averaging, median or corner method.
```
    *****************************************/

shrink_image_array(in_name, out_name,
    the_image, out_image,
    il1, ie1, ll1, le1, il2, ie2, ll2, le2,
    scale, method)
 char in_name[], out_name[], method[];
 int il1, ie1, ll1, le1,
    il2, ie2, ll2, le2, scale;
 short the_image[ROWS][COLS],
    out_image[ROWS][COLS];
{
   int A, B, a, b, count, factor,
        i, j, length, width;
   struct tiff_header_struct image_header;

    /*****************************************
```
Check the scale factor. If it is not a valid factor (2 or 4), then set it to 2.
```
    *****************************************/
```

```
factor = scale;
if(factor != 2 &&
   factor != 4) factor = 2;

create_file_if_needed(in_name, out_name, out_image);

read_tiff_header(in_name, &image_header);

/*******************************
            Corner method
*******************************/

if(method[0] == 'c' || method[0] == 'C'){
   count = 1;
   for(A=0; A<factor; A++){
     for(B=0; B<factor; B++){
       printf("\n shrinking by corner %3d of %3d",
           count++, factor*factor);
       if(image_header.image_length < ll1+A*ROWS ||
          image_header.image_width < le1+B*COLS)
          blank_image_array(the_image);
       else
          read_tiff_image(in_name, the_image,
                il1+A*ROWS, ie1+B*COLS,
                    ll1+A*ROWS, le1+B*COLS);
     for(i=0; i<ROWS/factor; i++){
       for(j=0; j<COLS/factor; j++){
          out_image[i+A*ROWS/factor][j+B*COLS/factor] =
             the_image[factor*i][factor*j];
             } /* ends loop over j */
           } /* ends loop over i */
         } /* ends loop over B */
       } /* ends loop over A */
     write_array_into_tiff_image(out_name, out_image,
                     il2, ie2, ll2, le2);
} /* ends corner method */
   /*******************************
              Average Method
   *******************************/

if(method[0] == 'a' || method[0] == 'A'){
   count = 1;
   for(A=0; A<factor; A++){
     for(B=0; B<factor; B++){
       printf("\n shrinking by average %3d of %3d",
             count++, factor*factor);
```

```
       if(image_header.image_length < ll1+A*ROWS ||
          image_header.image_width < le1+B*COLS)
          blank_image_array(the_image);
else
    read_tiff_image(in_name, the_image,
             il1+A*ROWS, ie1+B*COLS,
             ll1+A*ROWS, le1+B*COLS);
for(i=0; i<ROWS/factor; i++){
    for(j=0; j<COLS/factor; j++){
       out_image[i+A*ROWS/factor][j+B*COLS/factor] =
          average_pixel(the_image, factor, i, j);
       } /* ends loop over j */
     } /* ends loop over i */
     } /* ends loop over B */
  } /* ends loop over A */
  write_array_into_tiff_image(out_name, out_image,
                    il2, ie2, ll2, le2);
} /* ends average method */

  /***********************
                    Median Method
  ***********************/
if(method[0] == 'm' || method[0] == 'M'){
    count = 1;
    for(A=0; A<factor; A++){
     for(B=0; B<factor; B++){
      printf("\n shrinking by median %3d of %3d",
                 count++, factor*factor);
      if(image_header.image_length < ll1+A*ROWS ||
         image_header.image_width < le1+B*COLS)
         blank_image_array(the_image);
else
    read_tiff_image(in_name, the_image,
           il1+A*ROWS, ie1+B*COLS,
           ll1+A*ROWS, le1+B*COLS);
for(i=0; i<ROWS/factor; i++){
   for(j=0; j<COLS/factor; j++){
   out_image[i+A*ROWS/factor][j+B*COLS/factor] =
       median_pixel(the_image, factor, i, j);
        } /* ends loop over j */
      } /* ends loop over i */
```

```
    } /* ends loop over B */
  } /* ends loop over A */
  write_array_into_tiff_image(out_name, out_image,
  il2, ie2, ll2, le2);

    } /* ends median method */

} /* ends shrink_image_array */

    /***********************************************
```

average_pixelfunction calculates the average pixel value of a factor x factor array of pixels inside the the_image array. The coordinates i and j point to the upper left-hand corner of the small array.

```
    ***********************************************/

average_pixel(the_image, factor, i, j)
    int factor, i, j;
    short the_image[ROWS][COLS];
{
    int a, b, result = 0;
    for(a=0; a<factor; a++)
       for(b=0; b<factor; b++)
          result = result +
                the_image[factor*i+a][factor*j+a];
    result = result/(factor*factor);
    return(result);
} /* ends average_pixel */

    /***********************************************
```

median_pixel function calculates the median pixel value of a factor x factor array of pixels inside the the_image array. The coordinates i and j point to the upperleft hand corner of the small array.

```
    ***********************************************/

median_pixel(the_image, factor, i, j)
    int factor, i, j;
    short the_image[ROWS][COLS];
{
    int a, b, count, ff, result = 0;
    short *elements;
    ff = factor*factor;
    elements = (short *) malloc(ff * sizeof(short));
     count = 0;
    for(a=0; a<factor; a++){
      for(b=0; b<factor; b++){
         elements[count] =
```

```
                the_image[factor*i+a][factor*j+b];
             count++;
          }
       }
    result = median_of(elements, &ff);
    free(elements);
    return(result);
} /* ends median_pixel */

    /**********************************************
get_scaling_options function queries the user for the parameters needed to perform
scaling.
    **********************************************/

get_scaling_options(zoom_shrink, scale, method)
    int *scale;
    char method[], zoom_shrink[];
{
    int not_finished = 1, response;

 while(not_finished){
    printf("\n\t1. Zoom or Shrink is - %s",
            zoom_shrink);
    printf("\n\t2. Scale factor is %d", *scale);
    printf("\n\t3. Scaling Method is - %s", method);
    printf(
    "\n\t    Replication or Interpolation for Zooming"
    "\n\t    Averaging Median or Corner for Shrinking");
    printf("\n\n\tEnter choice (0 = no change) _\b");
    get_integer(&response);
    if(response == 0){
      not_finished = 0;
    }
    if(response == 1){
        printf("\nEnter Zoom or Shrink (z or s) __\b");
        gets(zoom_shrink);
  }
  if(response == 2){
    printf("\nEnter Scale Factor (2 or 4) __\b");
    get_integer(scale);
  }
if(response == 3){
    printf("\nEnter Scaling Method:"
        "Replication or Interpolation for Zooming"
        "\n"
```

"Averaging Median or Corner for Shrinking"
"\n\t__\b");
   gets(method);
   }
 } /* ends while not_finished */
} /* ends get_scaling_options */

```
   /*********************************************
blank_image_array function blanks out an image array by filling it with zeros.
   *********************************************/
blank_image_array(image)
    short image[ROWS][COLS];
{
    int i, j;
    for(i=0; i<ROWS; i++)
       for(j=0; j<COLS; j++)
           image[i][j] = 0;
} /* ends blank_image_array */
```

B.5 Rotation of an Image

```
   /*********************************************
The program rotates or flips an image in 90-degree increments.
   *********************************************/

short **the_image;
short **out_image;
main(argc, argv)
      int argc;
      char *argv[];
{
    char in_name[MAX_NAME_LENGTH], out_name[MAX_NAME_LENGTH];
    int type;
    long length1, width1;
    if(argc != 4){
    printf("\n\nusage: flip in-file out-file type");
    exit(0);
 }
    strcpy(in_name, argv[1]);
    strcpy(out_name, argv[2]);
    type = atoi(argv[3]);
       /* check input file */
    if(does_not_exist(in_name)){
      printf("\nERROR input file %s does not exist",
          in_name);
```

```
        exit(0);
    } /* ends if does_not_exist */
    create_image_file(in_name, out_name);
        /* allocate the image arrays */
    get_image_size(in_name, &length1, &width1);
    the_image = allocate_image_array(length1, width1);
    out_image = allocate_image_array(length1, width1);
    read_image_array(in_name, the_image);
    flip_image(the_image, out_image, type, length1, width1);
    write_image_array(out_name, out_image);
    free_image_array(the_image, length1);
    free_image_array(out_image, length1);
} /* ends main */
flip_image(the_image, out_image,
        type,
        rows, cols)
        int type;
        long cols, rows;
        short **the_image,
          **out_image;
    {
        int cd2, i, j, rd2;

    /*******************************************
Check the rotation_type. If it is not a valid value, set it to 1.
    ********************************************/

        if(type != 1 &&
          type != 2 &&
          type != 3 &&
          type != 4 &&
          type != 5) type = 1;

    /*******************************************
    Rotate the image array as desired.
    ********************************************/
    /*******************************************
        90-degree rotation
    ********************************************/

        if(type == 1 || type == 2 || type == 3){
            for(i=0; i<rows; i++){
                for(j=0; j<cols; j++)
                    out_image[j][cols-1-i] = the_image[i][j];
            } /* ends loop over i */
        } /* ends if type == 1 or 2 or 3 */
```

```
/*********************************************
      second 90-degree rotation
*********************************************/

if(type == 2 || type == 3){
    for(i=0; i<rows; i++)
        for(j=0; j<cols; j++)
            the_image[i][j] = out_image[i][j];
    for(i=0; i<rows; i++){
        for(j=0; j<cols; j++)
            out_image[j][cols-1-i] = the_image[i][j];
    } /* ends loop over i */
} /* ends if type == 2 or 3 */

/*********************************************
third 90-degree rotation
*********************************************/

if(type == 3){
    for(i=0; i<rows; i++)
        for(j=0; j<cols; j++)
            the_image[i][j] = out_image[i][j];
    for(i=0; i<rows; i++){
        for(j=0; j<cols; j++)
            out_image[j][cols-1-i] = the_image[i][j];
    } /* ends loop over i */
} /* ends if type == 3 */

/*********************************************
Flip the image array horizontally about the center vertical axis.
*********************************************/

if(type == 4){
    cd2 = cols/2;
    for(j=0; j<cd2; j++){
        for(i=0; i<rows; i++){
            out_image[i][cols-1-j] = the_image[i][j];
        } /* ends loop over i */
    } /* ends loop over j */

    for(j=cd2; j<cols; j++){
        for(i=0; i<rows; i++){
            out_image[i][cols-1-j] = the_image[i][j];
        } /* ends loop over i */
    } /* ends loop over j */
} /* ends if type == 4 */
```

```
/*******************************************
    Flip the image array vertically about the center horizontal axis.
    *******************************************/

    if(type == 5){
        rd2 = rows/2;
        for(i=0; i<rd2; i++){
            for(j=0; j<cols; j++){
                out_image[rows-1-i][j] = the_image[i][j];
            } /* ends loop over j */
        } /* ends loop over i */

        for(i=rd2; i<rows; i++){
            for(j=0; j<cols; j++){
                out_image[rows-1-i][j] = the_image[i][j];
            } /* ends loop over j */
        } /* ends loop over i */
    } /* ends if type == 5 */
} /* ends flip_image */
```

B.6 Geometric Transformations

```
/*******************************************
This file contains the main calling routine for geometric sub-routines.
    *******************************************/
    short **the_image;
    short **out_image;
    main(argc, argv)
        int argc;
        char *argv[];
    {
        char name1[80], name2[80], type[80];
        float theta, x_stretch, y_stretch,
            x_cross, y_cross;
        int bilinear;
        int x_control, y_control;
        long length, width;
        short m, n, x_displace, y_displace;

/**********************************
This program will use a different command line for each type of
call.
    **********************************/

    if(argc < 7){
    printf("\n\nNot enough parameters:");
```

```
printf("\n");
printf("\n     Two Operations:");
printf("\n        geometry rotate");
printf("\n\n     Examples:");
printf("\n");
printf("\n   geometry in out geometry angle");
printf(" x-displace y-displace");
printf("\n                 x-stretch y-stretch");
printf(" x-cross y-cross bilinear (1 or 0)");
printf("\n");
printf("\n geometry in out rotate angle m n");
printf(" bilinear (1 or 0)");
printf("\n");
exit(0);
}

/***********************************
Interpret the command line depending on the type of call.
***********************************/
if(strncmp(argv[3], "geometry", 3) == 0){
    strcpy(name1, argv[1]);
    strcpy(name2, argv[2]);
    strcpy(type, argv[3]);
    theta = atof(argv[4]);
    x_displace = atoi(argv[5]);
    y_displace = atoi(argv[6]);
    x_stretch = atof(argv[7]);
    y_stretch = atof(argv[8]);
    x_cross = atof(argv[9]);
    y_cross = atof(argv[10]);
    bilinear = atoi(argv[11]);
}
if(strncmp(argv[3], "rotate", 3) == 0){
    strcpy(name1, argv[1]);
    strcpy(name2, argv[2]);
    strcpy(type, argv[3]);
    theta = atof(argv[4]);
    m = atoi(argv[5]);
    n = atoi(argv[6]);
bilinear = atoi(argv[7]);
}
if(does_not_exist(name1)){
printf("\nERROR input file %s does not exist",
name1);
```

```
    exit(0);
    }
get_image_size(name1, &length, &width);
the_image = allocate_image_array(length, width);
out_image = allocate_image_array(length, width);
create_image_file(name1, name2);
read_image_array(name1, the_image);

/************************************
          Call the routines
************************************/

if(strncmp(type, "geometry", 3) == 0){
   geometry(the_image, out_image,
      theta, x_stretch, y_stretch,
      x_displace, y_displace,
      x_cross, y_cross,
      bilinear,
      length,
      width);
} /* ends if */
if(strncmp(type, "rotate", 3) == 0){
   arotate(the_image, out_image,
      theta, m, n, bilinear,
      length,
      width);
} /* ends if */
write_image_array(name2, out_image);
free_image_array(out_image, length);
free_image_array(the_image, length);
} /* ends main */
```

B.7 Spatial Domain Enhancement Filters

```
/*******************************************
* Define the filter masks.
*******************************************/
short lpf_filter_6[3][3] =
    {{0, 1, 0},
     {1, 2, 1},
     {0, 1, 0}};
short lpf_filter_9[3][3] =
    {{1, 1, 1},
     {1, 1, 1},
     {1, 1, 1}};
```

```
short lpf_filter_10[3][3] =
    { {1, 1, 1},
      {1, 2, 1},
      {1, 1, 1}};
short lpf_filter_16[3][3] =
    { {1, 2, 1},
      {2, 4, 2},
      {1, 2, 1}};
short lpf_filter_32[3][3] =
    { {1, 4, 1},
      {4, 12, 4},
      {1, 4, 1}};
short hpf_filter_1[3][3] =
    { { 0, -1, 0},
      {-1, 5, -1},
      { 0, -1, 0}};
short hpf_filter_2[3][3] =
    { {-1, -1, -1},
      {-1, 9, -1},
      {-1, -1, -1}};
short hpf_filter_3[3][3] =
    { { 1, -2, 1},
      {-2, 5, -2},
      { 1, -2, 1}};

/******************************************
filter_image function filters an image by using a single 3x3 mask.
******************************************/

    filter_image(the_image, out_image,
        rows, cols, bits_per_pixel,
        filter, type, low_high)
    int type;
    short filter[3][3],
        **the_image,
        **out_image;
    char low_high[];
    long rows, cols, bits_per_pixel;
{
    int a, b, d, i, j, k,
        length, max, sum, width;
    setup_filters(type, low_high, filter);
    d = type;
    if(type == 2 || type == 3) d = 1;
    max = 255;
```

```
   if(bits_per_pixel == 4)
      max = 16;
            /* Do convolution over image array */
   printf("\n");
   for(i=1; i<rows-1; i++){
      if( (i%10) == 0) printf("%d", i);
      for(j=1; j<cols-1; j++){
         sum = 0;
         for(a=-1; a<2; a++){
   for(b=-1; b<2; b++){
      sum = sum +
         the_image[i+a][j+b] *
         filter[a+1][b+1];
         }
      }
      sum                    = sum/d;
      if(sum < 0) sum = 0;
      if(sum > max) sum = max;
      out_image[i][j] = sum;

   } /* ends loop over j */
  } /* ends loop over i */
   fix_edges(out_image, 1, rows-1, cols-1);
 } /* ends filter_image */

   /*****************************************
```

median_filter function performs a median filter on an image using a size (3x3, 5x5, etc.) specified in the call.

```
   *****************************************/

   median_filter(the_image, out_image,
            rows, cols, size)
         int size;
         short **the_image,
               **out_i3mage;
         long rows, cols;

   {
      int a, b, count, i, j, k,
         length, sd2, sd2p1, ss, width;
      short *elements;
      sd2 = size/2;
      sd2p1 = sd2 + 1;
```

```
/***********************************************
Allocate the elements array large enough to hold size*size shorts.
***********************************************/

    ss  = size*size;
    elements = (short *) malloc(ss * sizeof(short));

    /**************************
            Loop over image array
    ************************/

printf("\n");
for(i=sd2; i<rows-sd2; i++){
    if( (i%10) == 0) printf("%d", i);
    for(j=sd2; j<cols-sd2; j++){
        count = 0;
        for(a=-sd2; a<sd2p1; a++){
            for(b=-sd2; b<sd2p1; b++){
                elements[count] = the_image[i+a][j+b];
                count++;
            }
        }
        out_image[i][j] = median_of(elements, &ss);
    } /* ends loop over j */
} /* ends loop over i */
free(elements);
fix_edges(out_image, sd2, rows-1, cols-1);
    } /* ends median_filter */

    /***********************************************
median_of function finds and returns the median value of the elements array. As a
side result, it also sorts the elements array.
***********************************************/

    median_of(elements, count)
            int *count;
            short elements[];
    {
        short median;
        fsort_elements(elements, count);
        median = elements[*count/2];
        return(median);
    } /* ends median_of */

    /*********************************************
low_pixel function replaces the pixel at the center of a 3x3, 5x5, etc. Area with
the min for that area.
*********************************************/
```

```
        low_pixel(the_image, out_image,
            rows, cols, size)
        int size;
        short **the_image,
            **out_image;
        long rows, cols;
{
        int a, b, count, i, j, k,
                length, sd2, sd2p1, ss, width;
        short *elements;
        sd2 = size/2;
        sd2p1 = sd2 + 1;

        /***********************************************
Allocate the elements array large enough to hold size*size shorts.
        ***********************************************/

        ss = size*size;
        elements = (short *) malloc(ss * sizeof(short));

        /*************************
            Loop over image array
        **************************/

        printf("\n");
        for(i=sd2; i<rows-sd2; i++){
            if( (i%10) == 0) printf("%d", i);
            for(j=sd2; j<cols-sd2; j++){
                count = 0;
                for(a=-sd2; a<sd2p1; a++){
                    for(b=-sd2; b<sd2p1; b++){
                        elements[count] = the_image[i+a][j+b];
                        count++;
                }
            }
            fsort_elements(elements, &ss);
            out_image[i][j] = elements[0];
    } /* ends loop over j */
} /* ends loop over i */

        free(elements);
        fix_edges(out_image, sd2, rows-1, cols-1);
} /* ends low_pixel */

        /*****************************************
high_pixel function replaces the pixel at the center of a 3x3, 5x5, etc. area with
the pixel value for that area.
        *****************************************/
```

```c
high_pixel(the_image, out_image,
      rows, cols, size)
   int size;
   short **the_image,
         **out_image;
   long rows, cols;
{
   int a, b, count, i, j, k,
      length, sd2, sd2p1, ss, width;
   short *elements;

   sd2 = size/2;
   sd2p1 = sd2 + 1;

/**********************************************
Allocate the elements array large enough to hold size*size shorts.
**********************************************/

   ss   = size*size;
   elements = (short *) malloc(ss * sizeof(short));

/**************************
   Loop over image array
**************************/
   printf("\n");
   for(i=sd2; i<rows-sd2; i++){
      if( (i%10) == 0) printf("%d", i);
      for(j=sd2; j<cols-sd2; j++){
            count = 0;
            for(a=-sd2; a<sd2p1; a++){
               for(b=-sd2; b<sd2p1; b++){
               elements[count] = the_image[i+a][j+b];
               count++;
            }
         }
         fsort_elements(elements, &ss);
         out_image[i][j] = elements[ss-1];
      } /* ends loop over j */
   } /* ends loop over i */

   free(elements);
   fix_edges(out_image, sd2, rows-1, cols-1);

} /* ends high_pixel */
```

```
/**********************************************
fsort_elements function performs a simple bubble sort on the elements from the
median filter.
**********************************************/

    fsort_elements(elements, count)
        int *count;
        short elements[];
    {
        int i, j;
        j = *count;
        while(j-- > 1){
            for(i=0; i<j; i++){
                if(elements[i] > elements[i+1])
                    fswap(&elements[i], &elements[i+1]);
            }
        }
    } /* ends fsort_elements */

/**********************************************
Fswap function swaps two shorts.
**********************************************/

    fswap(a, b)
  short *a, *b;
{
  short temp;
  temp = *a;
  *a = *b;
  *b = temp;
} /* ends swap */

/**********************************************
setup_filters function copies the filter mask values defined at the top of this file
into the filter array.
**********************************************/

setup_filters(filter_type, low_high, filter)
    char low_high[];
    int filter_type;
    short filter[3][3];
{
    int i, j;
if(low_high[0] == 'l' || low_high[0] =='L'){
    if(filter_type == 6){
        for(i=0; i<3; i++){
```

```
          for(j=0; j<3; j++){
             filter[i][j] = lpf_filter_6[i][j];
             }
          }
       } /* ends if filter_type == 6 */
if(filter_type == 9){
   for(i=0; i<3; i++){
      for(j=0; j<3; j++){
         filter[i][j] = lpf_filter_9[i][j];
         }
      }
} /* ends if filter_type == 9 */
if(filter_type == 10){
   for(i=0; i<3; i++){
      for(j=0; j<3; j++){
         filter[i][j] = lpf_filter_10[i][j];
         }
      }
} /* ends if filter_type == 10 */
if(filter_type == 16){
   for(i=0; i<3; i++){
      for(j=0; j<3; j++){
         filter[i][j] = lpf_filter_16[i][j];
         }
      }
} /* ends if filter_type == 16 */
if(filter_type == 32){
   for(i=0; i<3; i++){
      for(j=0; j<3; j++){
         filter[i][j] = lpf_filter_32[i][j];
         }
      }
   } /* ends if filter_type == 32 */
} /* ends low pass filter */
if(low_high[0] == 'h' || low_high[0] =='H'){
   if(filter_type == 1){
      for(i=0; i<3; i++){
         for(j=0; j<3; j++){
            filter[i][j] = hpf_filter_1[i][j];
         }
      }
} /* ends if filter_type == 1 */
```

```
if(filter_type == 2){
     for(i=0; i<3; i++){
         for(j=0; j<3; j++){
             filter[i][j] = hpf_filter_2[i][j];
         }
     }
} /* ends if filter_type == 2 */
if(filter_type == 3){
    for(i=0; i<3; i++){
         for(j=0; j<3; j++){
             filter[i][j] = hpf_filter_3[i][j];
             }
         }
       } /* ends if filter_type == 3 */
    } /* ends high pass filter */
} /* ends setup_filters */
```

B.8 Histogram Calculation

```
#define PRINT_WIDTH 80
#define FORMFEED '\014'

    /*****************************************
zero_histogram function clears or zeros a histogram array.
    *****************************************/

zero_histogram(histogram, gray_levels)
    int gray_levels;
    unsigned long histogram[];
{
    int i;
    for(i=0; i<gray_levels; i++)
        histogram[i] = 0;
} /* ends zero_histogram */

    /*****************************************
calculate_histogram function calculates the histogram for an input image array.
    *****************************************/

calculate_histogram(image, histogram, length, width)
    int length, width;
    short **image;
    unsigned long histogram[];
{
    long i,j;
    short k;
```

```
    for(i=0; i<length; i++){
        for(j=0; j<width; j++){
            k = image[i][j];
            histogram[k] = histogram[k] + 1;
            }
        }
} /* ends calculate_histogram */
```

```
    /*********************************************
smooth_histogram function smoothens the input histogram and returns it. It uses
a simple averaging scheme where each point in the histogram is replaced by the
average of itself and the two points on either side of it.
    *********************************************/
```

```
smooth_histogram(histogram, gray_levels)
    int gray_levels;
    unsigned long histogram[];
{
    int i;
    unsigned long new_hist[gray_levels];

    zero_histogram(new_hist, gray_levels);

new_hist[0] = (histogram[0] + histogram[1])/2;
new_hist[gray_levels] =
    (histogram[gray_levels] +
    histogram[gray_levels-1])/2;

for(i=1; i<gray_levels-1; i++){
    new_hist[i] = (histogram[i-1] +
        histogram[i] +
        histogram[i+1])/3;
}
for(i=0; i<gray_levels; i++)
    histogram[i] = new_hist[i];

} /* ends smooth_histogram */
```

```
    /*********************************************
perform_histogram_equalization function performs histogram equalization on the
input image array.
    *********************************************/
```

```
perform_histogram_equalization(image,
            histogram,
            gray_levels,
            new_grays,
            length,
```

```
        width)
    int gray_levels, new_grays;
    long length, width;
    short **image;
    unsigned long histogram[];
{
  int i,
     j,
     k;
unsigned long sum,
    sum_of_h[gray_levels];
double constant;
sum = 0;
for(i=0; i<gray_levels; i++){
    sum  = sum + histogram[i];
sum_of_h[i] = sum;
}
  /* constant = new # of gray levels div by area */
  constant = (float)(new_grays)/(float)(length*width);
  for(i=0; i<length; i++){
    for(j=0; j<width; j++){
       k = image[i][j];
       image[i][j] = sum_of_h[k] * constant;
     }
    }
} /* ends perform_histogram_equalization */

hist_long_clear_buffer(string)
    char string[];
{
    int i;
    for(i=0; i<300; i++)
        string[i] = ' ';
}
```

B.9 Image Segmentation

```
    /**************************************************
manual_threshold_segmentation function segments an image using thresholding
given the high and low values of the threshold by the calling routine.
    **************************************************/

manual_threshold_segmentation(in_name, out_name,
        the_image, out_image,
        il, ie, ll, le,
```

```
         hi, low, value, segment)
    char in_name[], out_name[];
    int il, ie, ll, le, segment;
    short hi, low, the_image[ROWS][COLS],
       out_image[ROWS][COLS], value;
{
    int length, width;
    struct tiff_header_struct image_header;

    create_file_if_needed(in_name, out_name, out_image);

    read_tiff_image(in_name, the_image, il, ie, ll, le);
    threshold_image_array(the_image, out_image,
             hi, low, value);
    if(segment == 1)
       grow(out_image, value);
    write_array_into_tiff_image(out_name, out_image,
             il, ie, ll, le);

} /* ends manual_threshold_segmentation */
    /**************************************************
```

peak_threshold_segmentation function segments an image using thresholding. It uses the histogram peaks to find the high and low values of the threshold.

```
    **************************************************/

peak_threshold_segmentation(in_name, out_name,
          the_image, out_image,
          il, ie, ll, le,
          value, segment)
    char in_name[], out_name[];
    int il, ie, ll, le, segment;
    short the_image[ROWS][COLS],
       out_image[ROWS][COLS], value;
{
    int    length, peak1, peak2, width;
    short  hi, low;
    struct tiff_header_struct image_header;
    unsigned long histogram[GRAY_LEVELS+1];

    create_file_if_needed(in_name, out_name, out_image);

    read_tiff_image(in_name, the_image, il, ie, ll, le);
    zero_histogram(histogram);
    calculate_histogram(the_image, histogram);
    smooth_histogram(histogram);
    find_peaks(histogram, &peak1, &peak2);
```

```
    peaks_high_low(histogram, peak1, peak2,
&hi, &low);
    threshold_image_array(the_image, out_image,
            hi, low, value);
    if(segment == 1)
       grow(out_image, value);
    write_array_into_tiff_image(out_name, out_image,
               il, ie, ll, le);
} /* ends peak_threshold_segmentation */

    /**********************************************
```

valley_threshold_segmentation function segments an image using thresholding. It uses the histogram valleys to find the high and low values of the threshold.

```
    **********************************************/
valley_threshold_segmentation(in_name, out_name,
                the_image, out_image,
                il, ie, ll, le,
                value, segment)
    char in_name[], out_name[];
    int il, ie, ll, le, segment;
    short the_image[ROWS][COLS],
       out_image[ROWS][COLS], value;
{
    int length, peak1, peak2, width;
    short hi, low;
    struct tiff_header_struct image_header;
    unsigned long histogram[GRAY_LEVELS+1];

    create_file_if_needed(in_name, out_name, out_image);

    read_tiff_image(in_name, the_image, il, ie, ll, le);
    zero_histogram(histogram);
    calculate_histogram(the_image, histogram);
    smooth_histogram(histogram);
    find_peaks(histogram, &peak1, &peak2);
    valley_high_low(histogram, peak1, peak2,
&hi, &low);
    threshold_image_array(the_image, out_image,
                hi, low, value);
    if(segment == 1)
       grow(out_image, value);
    write_array_into_tiff_image(out_name, out_image,
               il, ie, ll, le);
} /* ends valley_threshold_segmentation */
```

```
      /**************************************************
adaptive_threshold_segmentation function segments an image using thresholding. It
uses two passes to find the high and low values of the threshold. The first pass uses
the peaks of the histogram to find the high and low threshold values. It thresholds the
image using these highs/lows and calculates the means of the object and background.
Then these means are used as new peaks to calculate new high and low values.
      **************************************************/

adaptive_threshold_segmentation(in_name, out_name,
                the_image, out_image,
                il, ie, ll, le,
                value, segment)
    char in_name[], out_name[];
    int il, ie, ll, le, segment;
    short the_image[ROWS][COLS],
        out_image[ROWS][COLS], value;
{

    int length, peak1, peak2, width;
    short background, hi, low, object;
    struct tiff_header_struct image_header;
    unsigned long histogram[GRAY_LEVELS+1];

    create_file_if_needed(in_name, out_name, out_image);

    read_tiff_image(in_name, the_image, il, ie, ll, le);
    zero_histogram(histogram);
    calculate_histogram(the_image, histogram);
    smooth_histogram(histogram);
    find_peaks(histogram, &peak1, &peak2);
    peaks_high_low(histogram, peak1, peak2,
&hi, &low);
    threshold_and_find_means(the_image, out_image,
                hi, low, value,
&object, &background);
    peaks_high_low(histogram, object, background,
&hi, &low);
    threshold_image_array(the_image, out_image,
                hi, low, value);
    if(segment == 1)
        grow(out_image, value);
    write_array_into_tiff_image(out_name, out_image,
                        il, ie, ll, le);

} /* ends adaptive_threshold_segmentation */
```

```
/**************************************************
threshold_image_array function thresholds an input image array and produces a
binary output image array.
**************************************************/

threshold_image_array(in_image, out_image, hi, low, value)
    short hi, low, in_image[ROWS][COLS],
        out_image[ROWS][COLS], value;
{
    int counter = 0, i, j;
    for(i=0; i<ROWS; i++){
        for(j=0; j<COLS; j++){
            if(in_image[i][j] >= low &&
                in_image[i][j] <= hi){
                out_image[i][j] = value;
                counter++;
            }
            else
                out_image[i][j] = 0;
        } /* ends loop over j */
    } /* ends loop over i */
    printf("\n\tTIA> set %d points", counter);
} /* ends threshold_image_array */

    /**************************************************
threshold_and_find_means function thresholds an input image array and produces
a binary output image array.
**************************************************/

threshold_and_find_means(in_image, out_image, hi,
                low, value, object_mean,
                background_mean)
    short *background_mean, hi, low,
        in_image[ROWS][COLS], *object_mean,
        out_image[ROWS][COLS], value;
{
    int    counter = 0,
        i,
        j;
    unsigned long object = 0,
        background = 0;
    for(i=0; i<ROWS; i++){
        for(j=0; j<COLS; j++){
            if(in_image[i][j] >= low &&
                in_image[i][j] <= hi){
```

```
                out_image[i][j] = value;
                counter++;
                object = object + in_image[i][j];
            }
            else{
                out_image[i][j] = 0;
                background = background + in_image[i][j];
            }
        } /* ends loop over j */
    } /* ends loop over i */
    object = object/counter;
    background = background/((ROWS*COLS)-counter);
    *object_mean = (short)(object);
    *background_mean = (short)(background);
    printf("\n\tTAFM> set %d points", counter);
    printf("\n\tTAFM> object=%d background=%d",
        *object_mean, *background_mean);
} /* ends threshold_and_find_means */

    /************************************************
```

Grow function is an object detector. Its input is a binary image array containing 0's and value's. It searches through the image and connects the adjacent values.

```
        ***********************************************/

grow(binary, value)
    short binary[ROWS][COLS],
        value;
{
    char name[80];

    int first_call,
        i,
        j,
        object_found,
        pointer,
        pop_i,
        pop_j,
        stack_empty,
        stack_file_in_use;

    short g_label, stack[STACK_SIZE][2];

    g_label = 2;
    object_found = 0;
    first_call = 1;

    for(i=0; i<ROWS; i++){
```

```
    for(j=0; j<COLS; j++){

        stack_file_in_use = 0;
        stack_empty = 1;
        pointer = -1;

  /*********************************
Search for the first pixel of a region.
  *********************************/

if(binary[i][j] == value){
    label_and_check_neighbor(binary, stack,
        g_label, &stack_empty, &pointer,
        i, j, value, &stack_file_in_use,
&first_call);
    object_found = 1;
    } /* ends if binary[i]j] == value */

    while(stack_empty == 0){
    pop_i = stack[pointer][0]; /* POP */
    pop_j = stack[pointer][1]; /* OPERATION */
    --pointer;
    if(pointer <= 0){
        if(stack_file_in_use){
            pop_data_off_of_stack_file(
                            stack,
&pointer,
&stack_file_in_use);
        } /* ends if stack_file_in_use */
        else{
            pointer = 0;
            stack_empty = 1;
        } /* ends else stack file is
            not in use */
    } /* ends if point <= 0 */

    label_and_check_neighbor(binary,
                stack, g_label,
&stack_empty,
&pointer, pop_i,
                pop_j, value,
&stack_file_in_use,
&first_call);
        } /* ends while stack_empty == 0 */

        if(object_found == 1){
            object_found = 0;
```

```
          ++g_label;
      } /* ends if object_found == 1 */

   } /* ends loop over j */
  } /* ends loop over i */
  printf("\nGROW> found %d objects", g_label);

} /* ends grow */
```

```
/*********************************************
```
label_and_check_neighbors function labels a pixel with an object label and then checks the pixel's 8 neighbors. If any of the neigbors are set, then they are also labeled.
```
*********************************************/
```

```
label_and_check_neighbor(binary_image, stack,
                g_label, stack_empty,
                pointer, r, e, value,
                stack_file_in_use,
                first_call)
int e,
   *first_call,
   *pointer,
   r,
   *stack_empty,
   *stack_file_in_use;
short binary_image[ROWS][COLS],
   g_label,
   stack[STACK_SIZE][2],
   value;
{
   int already_labeled = 0,
       i, j;
   if (binary_image[r][e] == g_label)
      already_labeled = 1;

   binary_image[r][e] = g_label;

   /************************************
          Look at the 8 neighors of the point r,
   ************************************/

   for(i=(r-1); i<=(r+1); i++){
      for(j=(e-1); j<=(e+1); j++){

         if((i>=0) &&
            (i<=ROWS-1) &&
```

```
                (j>=0) &&
                (j<=COLS-1)){

           if(binary_image[i][j] == value){
             *pointer = *pointer + 1;
             stack[*pointer][0] = i; /* PUSH */
             stack[*pointer][1] = j; /* OPERATION */
             *stack_empty = 0;

           if(*pointer >= (STACK_SIZE -
                      STACK_FILE_LENGTH)){
             push_data_onto_stack_file(stack,
                      pointer, first_call);
             *stack_file_in_use = 1;
        } /* ends if *pointer >=
             STACK_SIZE - STACK_FILE_LENGTH*/

        } /* end of if binary_image == value */
       } /* end if i and j are on the image */
     } /* ends loop over i rows        */
   } /* ends loop over j columns       */
 } /* ends label_and_check_neighbors */
push_data_onto_stack_file(stack, pointer, first_call)
   int *first_call, *pointer;
   short stack[STACK_SIZE][2];
{
   char backup_file_name[MAX_NAME_LENGTH];
   FILE *backup_file_pointer, *stack_file_pointer;
   int diff, i;
   short holder[STACK_FILE_LENGTH][2];

   printf("\nSFO> Start of push_data_onto_stack");

   diff = STACK_SIZE - STACK_FILE_LENGTH;
   for(i=0; i<STACK_FILE_LENGTH; i++){
      holder[i][0] = stack[i][0];
      holder[i][1] = stack[i][1];
   }
   for(i=0; i<diff; i++){
      stack[i][0] = stack[i + STACK_FILE_LENGTH][0];
      stack[i][1] = stack[i + STACK_FILE_LENGTH][1];
   }
   for(i=diff; i<STACK_SIZE; i++){
      stack[i][0] = 0;
      stack[i][1] = 0;
   }
```

```
      *pointer = *pointer - STACK_FILE_LENGTH;
      if(*first_call == 1){

         *first_call = *first_call + 1;
         if((stack_file_pointer = fopen(STACK_FILE,"wb"))
                        == NULL)
           printf("\nSFO> Could not open stack file");
         else{
            /*printf("\n\nSFO> Writing to stack file");*/
            fwrite(holder, sizeof(holder),
               1, stack_file_pointer);
            fclose(stack_file_pointer);
         } /* ends else could not open stack_file */

} /* ends if *first_call == 1 */
else{ /* else stack file has been used already */
      strcpy(backup_file_name, STACK_FILE);
      strcat(backup_file_name, ".bak");
      if((backup_file_pointer =
         fopen(backup_file_name, "wb")) == NULL)
         printf("\nSFO> Could not open backup file");
      else{
         /*printf("\n\nSFO> Writing to backup file");*/
         fwrite(holder, sizeof(holder),
            1, backup_file_pointer);
         fclose(backup_file_pointer);
} /* ends else could not open backup_file */
append_stack_files(backup_file_name,
            STACK_FILE, holder);
copy_stack_files(backup_file_name,
            STACK_FILE, holder);

} /* ends else first_call != 1 */

printf("--- End of push_data_onto_stack");

} /* ends push_data_onto_stack_file */

pop_data_off_of_stack_file(stack, pointer,
               stack_file_in_use)
   int *pointer, *stack_file_in_use;
   short stack[STACK_SIZE][2];
{
   char backup_file_name[MAX_NAME_LENGTH];
   FILE *backup_file_pointer, *stack_file_pointer;
   int i;
   long write_counter;
```

```
    short holder[STACK_FILE_LENGTH][2],
        holder2[STACK_FILE_LENGTH][2];
    printf("\nSFO> Start of pop_data_off_of_stack");
    write_counter = 0;
    strcpy(backup_file_name, STACK_FILE);
    strcat(backup_file_name, ".bak");
    if( (stack_file_pointer =
        fopen(STACK_FILE, "rb")) == NULL)
        printf("\nSFO> Could not open stack file");
    else{
        /*printf("\n\nSFO> Reading from stack file");*/
        fread(holder, sizeof(holder),
            1, stack_file_pointer);

        backup_file_pointer =
            fopen(backup_file_name, "wb");
        while( fread(holder2, sizeof(holder2),
                1, stack_file_pointer) ){
            fwrite(holder2, sizeof(holder2),
                1, backup_file_pointer);
            ++write_counter;
        } /* ends while reading */
        if(write_counter > 0)
            *stack_file_in_use = 1;

        else
            *stack_file_in_use = 0;
        fclose(backup_file_pointer);
        fclose(stack_file_pointer);
} /* ends else could not open stack file */

copy_stack_files(backup_file_name,
        STACK_FILE, holder2);
for(i=0; i<STACK_FILE_LENGTH; i++){
    stack[i][0] = holder[i][0];
    stack[i][1] = holder[i][1];
}
*pointer = *pointer + STACK_FILE_LENGTH - 1;

    printf("--- End of pop_data_off_of_stack");
} /* ends pop_data_off_of_stack_file */

    append_stack_files(first_file, second_file, holder)
    char first_file[], second_file[];
    short holder[STACK_FILE_LENGTH][2];
{
```

```
  FILE *first, *second;
  int i;

  if((first = fopen(first_file, "r+b")) == NULL)
    printf("\n\nSFO> Cannot open file %s",
         first_file);
  if((second = fopen(second_file, "rb")) == NULL)
    printf("\n\nSFO> Cannot open file %s",
          second_file);
  fseek(first, 0L, 2);
  fseek(second, 0L, 0);
  while(fread(holder, sizeof(holder), 1, second) ){
    fwrite(holder, sizeof(holder), 1, first);
  } /* ends while reading */
  fclose(first);
  fclose(second);
} /* ends append_stack_files */
copy_stack_files(first_file, second_file, holder)
  char first_file[], second_file[];
  short holder[STACK_FILE_LENGTH][2];
{
  FILE *first, *second;
  int i;

  if( (first = fopen(first_file, "rb")) == NULL)
    printf("\n\nSFO> Cannot open file %s",
       first_file);

  if( (second = fopen(second_file, "wb")) == NULL)
    printf("\n\nSFO> Cannot open file %s",
       second_file);
  fseek(first, 0L, 0);

  while( fread(holder, sizeof(holder), 1, first) ){
     fwrite(holder, sizeof(holder), 1, second);
  } /* ends while reading */

  fclose(first);
  fclose(second);
} /* ends copy_stack_files */
    /*******************************************
```

find_peaks function looks through the histogram array and finds the two highest peaks.

```
    *******************************************/
```

find_peaks(histogram, peak1, peak2)

```
    unsigned long histogram[];
    int *peak1, *peak2;
{
    int distance[PEAKS], peaks[PEAKS][2];
    int i, j=0, max=0, max_place=0;

    for(i=0; i<PEAKS; i++){
        distance[i] = 0;
        peaks[i][0] = -1;
        peaks[i][1] = -1;
}

for(i=0; i<=GRAY_LEVELS; i++){
    max = histogram[i];
    max_place = i;
    insert_into_peaks(peaks, max, max_place);
} /* ends loop over i */

for(i=1; i<PEAKS; i++){
    distance[i] = peaks[0][1] - peaks[i][1];
    if(distance[i] < 0)
        distance[i] = distance[i]*(-1);
}
*peak1 = peaks[0][1];
for(i=PEAKS-1; i>0; i--)
  if(distance[i] > PEAK_SPACE) *peak2 = peaks[i][1];
} /* ends find_peaks */

    /*******************************************
```

insert_into_peaks function takes a value and its place in the histogram and inserts
them into a peaks array. This helps us rank the peaks in the histogram.

```
    *******************************************/

insert_into_peaks(peaks, max, max_place)
    int max, max_place, peaks[PEAKS][2];
{
    int i, j;
        /* first case */
    if(max > peaks[0][0]){
        for(i=PEAKS-1; i>0; i--){
            peaks[i][0] = peaks[i-1][0];
            peaks[i][1] = peaks[i-1][1];
        }
        peaks[0][0] = max;
        peaks[0][1] = max_place;
} /* ends if */
```

```
    /* middle cases */
for(j=0; j<PEAKS-3; j++){
   if(max < peaks[j][0] && max > peaks[j+1][0]){
      for(i=PEAKS-1; i>j+1; i--){
         peaks[i][0] = peaks[i-1][0];
         peaks[i][1] = peaks[i-1][1];
      }
      peaks[j+1][0] = max;
      peaks[j+1][1] = max_place;
   } /* ends if */
} /* ends loop over j */
/* last case */
if(max < peaks[PEAKS-2][0] &&
max > peaks[PEAKS-1][0]){
peaks[PEAKS-1][0] = max;
peaks[PEAKS-1][1] = max_place;
 } /* ends if */

} /* ends insert_into_peaks */

   /*********************************************
```

peaks_high_low function uses the histogram array and the peaks to find the best
high and low threshold values for the threshold function.
```
   *********************************************/

peaks_high_low(histogram, peak1, peak2, hi, low)
   int peak1, peak2;
   short *hi, *low;
   unsigned long histogram[];
{
 int i, mid_point;
   unsigned long sum1 = 0, sum2 = 0;
   if(peak1 > peak2)
      mid_point = ((peak1 - peak2)/2) + peak2;
   if(peak1 < peak2)
      mid_point = ((peak2 - peak1)/2) + peak1;
   for(i=0; i<mid_point; i++)
      sum1 = sum1 + histogram[i];
   for(i=mid_point; i<=GRAY_LEVELS; i++)
      sum2 = sum2 + histogram[i];
   if(sum1 >= sum2){
      *low = mid_point;
      *hi = GRAY_LEVELS;

   }
   else{
```

```
      *low = 0;
        *hi = mid_point;
    }
} /* ends peaks_high_low */
```

        /***

valley_high_low function uses the histogram array and the valleys to find the best high and low threshold values for the threshold function.
        ***/

```
valley_high_low(histogram, peak1, peak2, hi, low)
    int peak1, peak2;
    short *hi, *low;
    unsigned long histogram[];
{
    int i, valley_point;
    unsigned long sum1 = 0, sum2 = 0;

    find_valley_point(histogram, peak1, peak2,
&valley_point);
    /*printf("\nVHL> valley point is %d",
        valley_point);*/

 for(i=0; i<valley_point; i++)
    sum1 = sum1 + histogram[i];
 for(i=valley_point; i<=GRAY_LEVELS; i++)
    sum2 = sum2 + histogram[i];

 if(sum1 >= sum2){
    *low = valley_point;
    *hi = GRAY_LEVELS;
 }
 else{
    *low = 0;
    *hi = valley_point;
 }

} /* ends valley_high_low */
```

        /***

find_valley_point function finds the low point of the valley between two peaks in a histogram. It starts at the lowest peak and works its way up to the highest peak. Along the way, it looks at each point in the histogram and inserts them into a list of points. When done, it has the location of the smallest histogram point - that is the valley point. The deltas array holds the delta value in the first place and its location in the second place.
        ***/

```
find_valley_point(histogram, peak1,
            peak2, valley_point)
    int peak1, peak2, *valley_point;
    unsigned long histogram[];
{
int deltas[PEAKS][2], delta_hist, i;

for(i=0; i<PEAKS; i++){
    deltas[i][0] = 10000;
    deltas[i][1] = -1;
}

if(peak1 < peak2){
    for(i=peak1+1; i<peak2; i++){
        delta_hist = (int)(histogram[i]);
        insert_into_deltas(deltas, delta_hist, i);
    } /* ends loop over i */
} /* ends if peak1 < peak2 */

if(peak2 < peak1){
    for(i=peak2+1; i<peak1; i++){
        delta_hist = (int)(histogram[i]);
        insert_into_deltas(deltas, delta_hist, i);
    } /* ends loop over i */
} /* ends if peak2 < peak1 */

*valley_point = deltas[0][1];

} /* ends find_valley_point */
```

```
    /*********************************************
```

insert_into_deltas function inserts histogram deltas into a deltas array. The smallest
delta will be at the top of the array.
```
    *********************************************/
```

```
insert_into_deltas(deltas, value, place)
    int value, place, deltas[PEAKS][2];
{
    int i, j;
        /* first case */
    if(value < deltas[0][0]){
        for(i=PEAKS-1; i>0; i--){
            deltas[i][0] = deltas[i-1][0];
            deltas[i][1] = deltas[i-1][1];
        }
        deltas[0][0] = value;
        deltas[0][1] = place;
```

```
      } /* ends if */

      /* middle cases */
  for(j=0; j<PEAKS-3; j++){
     if(value > deltas[j][0] &&
        value < deltas[j+1][0]){
        for(i=PEAKS-1; i>j+1; i--){
           deltas[i][0] = deltas[i-1][0];
           deltas[i][1] = deltas[i-1][1];
        }
        deltas[j+1][0] = value;
        deltas[j+1][1] = place;
     } /* ends if */

  } /* ends loop over j */

      /* last case */
    if(value > deltas[PEAKS–2][0] &&
      value < deltas[PEAKS-1][0]){
      deltas[PEAKS-1][0] = value;
      deltas[PEAKS-1][1] = place;
    } /* ends if */

} /* ends insert_into_deltas */

/*******************************************
```

get_segmentation_options function interacts with the user in order to obtain the options for image segmentation.

```
*********************************************/

get_segmentation_options(method, hi, low, value)
    char method[];
    short *hi, *low, *value;
{
    int i, not_finished = 1, response;
    while(not_finished){
        printf(
           "\n\nThe image segmentation options are:\n");
        printf("\n\t1. Method is %s", method);
        printf("\n\t (options are manual peaks");
        printf( " valleys adapative)");
        printf("\n\t2. Value is %d", *value);
        printf("\n\t3. Hi is %d", *hi);
        printf("\n\t4. Low is %d", *low);
        printf("\n\t Hi and Low needed only for");
        printf( " manual method");
        printf("\n\nEnter choice (0 = no change):_\b");
```

```
        get_integer(&response);
        if(response == 0)
            not_finished = 0;
        if(response == 1){
            printf("\nEnter method (options are:");
            printf(" manual peaks valleys adaptive)\n\t");
            gets(method);
        }
        if(response == 2){
            printf("\nEnter value: ___\b\b\b");
            get_short(value);
        }
        if(response == 3){
            printf("\nEnter hi: ___\b\b\b");
            get_short(hi);
        }
        if(response == 4){
            printf("\nEnter low: ___\b\b\b");
            get_short(low);
        }
  } /* ends while not_finished */
} /* ends get_segmentation_options */

    /*********************************************
get_threshold_options function interacts with the user in order to obtain the options
for image threshold.
    *********************************************/

get_threshold_options(method, hi, low, value)
        char method[];
        short *hi, *low, *value;
{
    int i, not_finished = 1, response;

  while(not_finished){
    printf("\n\nThe image threshold options are:\n");
    printf("\n\t1. Method is %s", method);
    printf("\n\t (options are manual peaks");
    printf(" valleys adapative)");
    printf("\n\t2. Value is %d", *value);
    printf("\n\t3. Hi is %d", *hi);
    printf("\n\t4. Low is %d", *low);
    printf("\n\t Hi and Low needed only for");
    printf(" manual method");
    printf("\n\nEnter choice (0 = no change):_\b");
```

```
        get_integer(&response);
        if(response == 0)
            not_finished = 0;
        if(response == 1){
        printf("\nEnter method (options are:");
        printf(" manual peaks valleys adaptive)\n\t");
        gets(method);
    }
    if(response == 2){
        printf("\nEnter value: ___\b\b\b");
        get_short(value);
    }
    if(response == 3){
        printf("\nEnter hi: ___\b\b\b");
        get_short(hi);
    }
    if(response == 4){
        printf("\nEnter low: ___\b\b\b");
        get_short(low);
    }
    } /* ends while not_finished */
} /* ends get_threshold_options */

    /******************************************

find_cutoff_point function looks at a histogram and sets a cut-off point at a given
percentage of pixels. For example, if percent=0.6,
start at 0 in the histogram and count up until you've hit 60% of the pixels. Then
stop and return that pixel value.
    ******************************************/

find_cutoff_point(histogram, percent, cutoff)
    unsigned long histogram[];
    float percent;
    short *cutoff;
{
    float fd, fsum, sum_div;
    int i, looking;
    long lc, lr, num=0, sum=0;
    sum = 0;
i = 0;
lr = (long)(ROWS);
lc = (long)(COLS);
num = lr*lc;
fd = (float)(num);
while(looking){
```

```
    fsum = (float)(sum);
    sum_div = fsum/fd;
    if(sum_div >= percent)
        looking = 0;
    else
        sum = sum + histogram[i++];
 } /* ends while looking */

 if(i >= 256) i = 255;
 *cutoff = i;
 printf("\nCutoff is %d sum=%ld", *cutoff, sum);
 } /* ends find_cutoff_point */

    /*******************************************
edge_region function segments an image by growing regions inside of edges.
    *******************************************/

edge_region(in_name, out_name, the_image, out_image,
        il, ie, ll, le, edge_type, min_area,
        max_area, diff, percent, set_value,
        erode)
 char    in_name[], out_name[];
 float   percent;
 int     edge_type, il, ie, ll, le;
 short      diff, erode,
           max_area, min_area,
           set_value,
           the_image[ROWS][COLS],
           out_image[ROWS][COLS];
{

 int a, b, count, i, j, k,
        length, width;
 short cutoff;
 struct tiff_header_struct image_header;
 unsigned long histogram[GRAY_LEVELS+1];
 create_file_if_needed(in_name, out_name, out_image);
 read_tiff_image(in_name, the_image, il, ie, ll, le);
    if(edge_type == 1 ||
    edge_type == 2 ||
    edge_type == 3)
    detect_edges(in_name, out_name, the_image,
        out_image, il, ie, ll, le,
        edge_type, 0, 0);
 if(edge_type == 4){
    quick_edge(in_name, out_name, the_image,
```

```
                  out_image, il, ie, ll, le,
                  0, 0);
   } /* ends if 4 */
   if(edge_type == 5){
      homogeneity(in_name, out_name, the_image,
         out_image, il, ie, ll, le,
         0, 0);
   } /* ends if 5 */
   if(edge_type == 6){
      difference_edge(in_name, out_name, the_image,
         out_image, il, ie, ll, le,
         0, 0);
   } /* ends if 6 */
   if(edge_type == 7){
      contrast_edge(in_name, out_name, the_image,
            out_image, il, ie, ll, le,
            0, 0);
   } /* ends if 7 */
   if(edge_type == 8){
      gaussian_edge(in_name, out_name, the_image,
         out_image, il, ie, ll, le,
         3, 0, 0);
   } /* ends if 8 */
   if(edge_type == 10){
      range(in_name, out_name, the_image,
         out_image, il, ie, ll, le,
         3, 0, 0);
   } /* ends if 10 */
   if(edge_type == 11){
      variance(in_name, out_name, the_image,
         out_image, il, ie, ll, le,
         0, 0);
   } /* ends if 11 */
/**write_array_into_tiff_image("f:e1.tif", out_image,
il, ie, ll, le);**/
      /* copy out_image to the_image */
   for(i=0; i<ROWS; i++)
      for(j=0; j<COLS; j++)
            the_image[i][j] = out_image[i][j];

      /****************************
Threshold the edge detector output at a given percent. This eliminates the weak
edges.
      ****************************/
```

```
    zero_histogram(histogram);
    calculate_histogram(the_image, histogram);
    find_cutoff_point(histogram, percent, &cutoff);
    threshold_image_array(the_image, out_image,
                255, cutoff, set_value);
/**write_array_into_tiff_image("f:e2.tif", out_image,
il, ie, ll, le);**/

    if(erode != 0){
       /* copy out_image to the_image */
      for(i=0; i<ROWS; i++)
        for(j=0; j<COLS; j++)
            the_image[i][j] = out_image[i][j];
      erode_image_array(the_image, out_image,
                set_value, erode);
    } /* ends if erode */

/**write_array_into_tiff_image("f:e3.tif", out_image,
il, ie, ll, le);**/
    for(i=0; i<ROWS; i++)
      for(j=0; j<COLS; j++)
        if(out_image[i][j] == set_value)
            out_image[i][j] = FORGET_IT;
      for(i=0; i<ROWS; i++)
        for(j=0; j<COLS; j++)
            the_image[i][j] = out_image[i][j];
      pixel_grow(the_image, out_image, diff,
                min_area, max_area);
      write_array_into_tiff_image(out_name, out_image,
                il, ie, ll, le);
} /* ends edge_region */

    /******************************************
gray_shade_region function segments an image by growing regions based only
on gray shade.
    *******************************************/

gray_shade_region(in_name, out_name, the_image,
    out_image, il, ie, ll, le,
    diff, min_area, max_area)
char in_name[], out_name[];
int il, ie, ll, le;
short the_image[ROWS][COLS],
    out_image[ROWS][COLS],
    diff, min_area, max_area;
```

```
{
   int a, b, big_count, count, i, j, k, l,
        not_finished, length, width;
   short temp[3][3];
   struct tiff_header_struct image_header;
   create_file_if_needed(in_name, out_name, out_image);
   read_tiff_image(in_name, the_image, il, ie, ll, le);
   pixel_grow(the_image, out_image, diff,
        min_area, max_area);
   write_array_into_tiff_image(out_name, out_image,
                il, ie, ll, le);
} /* ends gray_shade_region */
```

/**

edge_gray_shade_region function segments an image by growing gray shade regions inside of edges. It combines the techniques of the edge_region and gray_ shade_region functions.

***/

```
edge_gray_shade_region(in_name, out_name, the_image,
        out_image, il, ie, ll, le, edge_type,
        min_area, max_area, diff, percent,
        set_value, erode)
   char in_name[], out_name[];
   float percent;
   int    edge_type, il, ie, ll, le;
 short    diff, erode,
           max_area, min_area,
           set_value,
           the_image[ROWS][COLS],
           out_image[ROWS][COLS];
{
   int a, b, count, i, j, k,
        length, width;
   short cutoff;
   struct tiff_header_struct image_header;
   unsigned long histogram[GRAY_LEVELS+1];
   create_file_if_needed(in_name, out_name, out_image);
   read_tiff_image(in_name, the_image, il, ie, ll, le);
   if(edge_type == 1 ||
     edge_type == 2 ||
     edge_type == 3)
     detect_edges(in_name, out_name, the_image,
        out_image, il, ie, ll, le,
        edge_type, 0, 0);
```

```
if(edge_type == 4){
    quick_edge(in_name, out_name, the_image,
        out_image, il, ie, ll, le,
        0, 0);
} /* ends if 4 */
if(edge_type == 5){
    homogeneity(in_name, out_name, the_image,
        out_image, il, ie, ll, le,
        0, 0);
} /* ends if 5 */
if(edge_type == 6){
    difference_edge(in_name, out_name, the_image,
        out_image, il, ie, ll, le,
        0, 0);
} /* ends if 6 */
if(edge_type == 7){
    contrast_edge(in_name, out_name, the_image,
        out_image, il, ie, ll, le,
        0, 0);
} /* ends if 7 */
if(edge_type == 8){
    gaussian_edge(in_name, out_name, the_image,
        out_image, il, ie, ll, le,
        3, 0, 0);
} /* ends if 8 */
if(edge_type == 10){
    range(in_name, out_name, the_image,
        out_image, il, ie, ll, le,
        3, 0, 0);
} /* ends if 10 */
if(edge_type == 11){
        variance(in_name, out_name, the_image,
        out_image, il, ie, ll, le,
        0, 0);
} /* ends if 11 */
/**write_array_into_tiff_image("f:e1.tif", out_image,
il, ie, ll, le);**/

    /* copy out_image to the_image */
    for(i=0; i<ROWS; i++)
        for(j=0; j<COLS; j++)
            the_image[i][j] = out_image[i][j];
    zero_histogram(histogram);
    calculate_histogram(the_image, histogram);
```

```
        find_cutoff_point(histogram, percent, &cutoff);
        threshold_image_array(the_image, out_image,
                      255, cutoff, set_value);
/**write_array_into_tiff_image("f:e2.tif", out_image,
il, ie, ll, le);**/

    if(erode != 0){
        /* copy out_image to the_image */
        for(i=0; i<ROWS; i++)
            for(j=0; j<COLS; j++)
                the_image[i][j] = out_image[i][j];
        erode_image_array(the_image, out_image,
                      set_value, erode);
    } /* ends if erode */
/**write_array_into_tiff_image("f:e3.tif", out_image,
il, ie, ll, le);**/

        /*****************************
            Read the original gray shade image back into the_image.
        *****************************/
    read_tiff_image(in_name, the_image, il, ie, ll, le);
        /*****************************
Overlay the edge values on top of the original image by setting them to FORGET_IT
so the region growing will not use those points.
        *****************************/

      for(i=0; i<ROWS; i++)
          for(j=0; j<COLS; j++)
              if(out_image[i][j] == set_value)
                  the_image[i][j] = FORGET_IT;
/**write_array_into_tiff_image("f:e4.tif", the_image,
il, ie, ll, le);**/
        pixel_grow(the_image, out_image, diff,
            min_area, max_area);

        write_array_into_tiff_image(out_name, out_image,
                      il, ie, ll, le);
} /* ends edge_gray_shade_region */
        /************************************************
pixel_grow function grows regions. It adds pixels to a growing region only if the
pixel is close enough to the average gray level of that region.
        ************************************************/
pixel_grow(input, output, diff, min_area, max_area)
    short input[ROWS][COLS],
        output[ROWS][COLS],
```

```
                max_area,
                min_area,
                diff;
{
    char name[80];
    int count,
        first_call,
        i,
        ii,
        j,
        jj,
        object_found,
        pointer,
        pop_i,
        pop_j,
        stack_empty,
        stack_file_in_use;
    short g_label, target, sum, stack[STACK_SIZE][2];
    for(i=0; i<ROWS; i++)
      for(j=0; j<COLS; j++)
        output[i][j] = 0;
    g_label = 2;
    object_found = 0;
    first_call = 1;

        /***********************************
                the process of growing regions.
        ***********************************/
      for(i=0; i<ROWS; i++){
if( (i%4) == 0) printf("\n");
printf("-i=%3d label=%3d", i, g_label);
        for(j=0; j<COLS; j++){
            target = input[i][j];
            sum = target;
            count = 0;
            stack_file_in_use = 0;
            stack_empty = 1;
            pointer = -1;

        /********************************
        Search for the first pixel of a region.
        ********************************/

            if(input[i][j] != FORGET_IT &&
                is_close(input[i][j], target, diff) &&
```

```
            output[i][j] == 0){
            pixel_label_and_check_neighbor(input,
                        output, &target, &sum,
&count, stack, g_label,
&stack_empty, &pointer,
                i, j, diff,
&stack_file_in_use,
&first_call);
        object_found = 1;
      } /* ends if is_close */
 while(stack_empty == 0){
    pop_i = stack[pointer][0]; /* POP */
    pop_j = stack[pointer][1]; /* OPERATION */
    --pointer;
    if(pointer <= 0){
       if(stack_file_in_use){
          pop_data_off_of_stack_file(
                stack,
&pointer,
&stack_file_in_use);
          } /* ends if stack_file_in_use */
          else{
             pointer = 0;
             stack_empty = 1;
          } /* ends else stack file is
             not in use */
       } /* ends if point <= 0 */
       pixel_label_and_check_neighbor(input,
                output, &target, &sum,
&count, stack, g_label,
&stack_empty, &pointer,
                pop_i, pop_j,
                diff, &stack_file_in_use,
&first_call);
    } /* ends while stack_empty == 0 */

    if(object_found == 1){
 object_found = 0;

   /*********************************
The object must be in the size constraints given by min_area and max_area
   *********************************/

   if(count >= min_area &&
      count <= max_area)
```

```
      ++g_label;

      /*********************************
                 Remove the object from the output.
      *********************************/
  else{
    for(ii=0; ii<ROWS; ii++){
      for(jj=0; jj<COLS; jj++){
        if(output[ii][jj] == g_label){
          output[ii][jj] = 0;
          input[ii][jj] = FORGET_IT;
          } /* ends if output == g_label */
        } /* ends loop over jj */
      } /* ends loop over ii */
    } /* ends else remove object */
  } /* ends if object_found == 1 */

  } /* ends loop over j */
} /* ends loop over i */
printf("\nGROW> found %d objects", g_label);

} /* ends pixel_grow */

      /*********************************************
pixel_label_and_check_neighbors function labels a pixel with an object label and
then checks the pixel's 8 neighbors. If any of the neigbors are set, then they are
also labeled.
      *********************************************/

pixel_label_and_check_neighbor(input_image,
                 output_image, target,
                 sum, count, stack,
                 g_label, stack_empty,
                 pointer, r, e, diff,
                 stack_file_in_use,
                 first_call)
int *count,
    e,
    *first_call,
    *pointer,
    r,
    *stack_empty,
    *stack_file_in_use;
    short input_image[ROWS][COLS],
    output_image[ROWS][COLS],
    g_label,
```

```
      *sum,
      *target,
      stack[STACK_SIZE][2],
      diff;
{
   int already_labeled = 0,
     i, j;
   if (output_image[r][e] != 0)
     already_labeled = 1;
   output_image[r][e] = g_label;
   *count = *count + 1;
   if(*count > 1){
     *sum = *sum + input_image[r][e];
     *target = *sum / *count;
}

   /***************************************
   Look at the 8 neighors of the point r,e.
   ***************************************/

for(i=(r-1); i<=(r+1); i++){
   for(j=(e-1); j<=(e+1); j++){

     if((i>=0) &&
        (i<=ROWS-1) &&
        (j>=0) &&
        (j<=COLS-1)){

       if( input_image[i][j] != FORGET_IT &&
         is_close(input_image[i][j],
              *target, diff)        &&
       output_image[i][j] == 0){
       *pointer = *pointer + 1;
       stack[*pointer][0] = i; /* PUSH */
       stack[*pointer][1] = j; /* OPERATION */
       *stack_empty = 0;
       if(*pointer >= (STACK_SIZE -
              STACK_FILE_LENGTH)){
       push_data_onto_stack_file(stack,
              pointer, first_call);
           *stack_file_in_use = 1;
         } /* ends if *pointer >=
             STACK_SIZE - STACK_FILE_LENGTH*/
             } /* ends if is_close */
           } /* end if i and j are on the image */
         } /* ends loop over i rows */
```

```
    } /* ends loop over j columns */
} /* ends pixel_label_and_check_neighbors */

    /******************************************
is_close function tests to see if two pixel values are close enough together. It uses
the delta parameter to make this judgement.
    *******************************************/

is_close(a, b, delta)
    short a, b, delta;
{
    int result = 0;
    short diff;
    diff = a-b;
    if(diff < 0) diff = diff*(-1);
    if(diff < delta)
        result = 1;
    return(result);
} /* ends is_close */

    /******************************************
erode_image_array function erodes pixels. If a pixel equals value and has more
than threshold neighbors equal to 0, then set that pixel in the output to 0.
    *******************************************/
erode_image_array(the_image, out_image,
        value, threshold)
    short the_image[ROWS][COLS],
        out_image[ROWS][COLS],
        threshold,
        value;
{
    int a, b, count, i, j, k,
        length, width;

    for(i=0; i<ROWS; i++)
      for(j=0; j<COLS; j++)
        out_image[i][j] = the_image[i][j];
    printf("\n");
    for(i=1; i<ROWS-1; i++){
 if( (i%10) == 0) printf("%3d", i);
 for(j=1; j<COLS-1; j++){
  if(the_image[i][j] == value){
     count = 0;
     for(a=-1; a<=1; a++){
        for(b=-1; b<=1; b++){
```

```
            if(the_image[i+a][j+b] == 0)
                count++;
                } /* ends loop over b */
            } /* ends loop over a */
            if(count > threshold) out_image[i][j] = 0;
        } /* ends if the_image == value */
    } /* ends loop over j */
  } /* ends loop over i */

} /* ends erode_image_array */
  /*****************************************
```

get_edge_region_options function interacts with the user to get the options needed to call the edge and region based segmentation routines.

```
      *****************************************/

get_edge_region_options(method, edge_type,
    min_area, max_area, set_value,
    diff, percent, erode)
char method[];
float *percent;
int *edge_type;
short *diff, *erode,
    *min_area, *max_area,
    *set_value;
{
int not_finished = 1, response;
while(not_finished){
      printf("\n\nEdge Region Segmentation Options:");
      printf("\n\t1.  Method is %s", method);
      printf("\n\t    Recall: Edge, Gray shade,"
              "Combination");
      printf("\n\t2.  Edge type is %d", *edge_type);
      printf("\n\t    Recall: ");
      printf("\n\t     1=Prewitt 2=Kirsch");
      printf("\n\t     3=Sobel 4=quick");
      printf("\n\t     5=homogeneity 6=difference");
      printf("\n\t     7=contrast 8=gaussian");
      printf("\n\t     10=range 11=variance");
      printf("\n\t3.  Min area is %d", *min_area);
      printf("\n\t4.  Max area is %d", *max_area);
      printf("\n\t5. Set value is %d", *set_value);
      printf("\n\t6. Difference value is %d", *diff);
      printf("\n\t7. Threshold percentage is %f",
                *percent);
      printf("\n\t8. Erode is %d", *erode);
```

```
            printf("\n\nEnter choice (0 = no change) _\b");
            get_integer(&response);
            if(response == 0){
                not_finished = 0;
            }
    if(response == 1){
        printf("\n\t Recall: Edge, Gray shade,"
                    "Combination");
        printf("\n\t>");
        gets(method);
    }
    if(response == 2){
        printf("\n\t    Recall:");
        printf("\n\t    1=Prewitt        2=Kirsch");
        printf("\n\t    3=Sobel          4=quick");
        printf("\n\t    5=homogeneity 6=difference");
        printf("\n\t    7=contrast       8=gaussian");
        printf("\n\t    10=range         11=variance");
        printf("\n\t__\b");
        get_integer(edge_type);
    }
    if(response == 3){
        printf("\nEnter min area:__\b\b");
        get_integer(min_area);
    }
    if(response == 4){
        printf("\nEnter max area:__\b\b");
        get_integer(max_area);
    }
    if(response == 5){
        printf("\nEnter set value:__\b\b");
        get_integer(set_value);
    }
    if(response == 6){
        printf("\nEnter difference:__\b\b");
        get_integer(diff);
    }
    if(response == 7){
        printf("\nEnter threshold percentage:__\b\b");
        get_float(percent);
    }
    if(response == 8){
        printf("\nEnter erode:__\b\b");
        get_integer(erode);
```

```
  }
 } /* ends while not_finished */
 } /* ends get_edge_region_options */
```

B.10 Edge Detection Algorithms

```
 short quick_mask[3][3] = {
     {-1, 0, -1},
     { 0, 4, 0},
     {-1, 0, -1} };

     /*************************
     * Directions for the masks
     * 3 2 1
     * 4 x 0
     * 5 6 7
     *
     *************************/

 /* masks for kirsch operator */
 short kirsch_mask_0[3][3] = {
         { 5, 5, 5},
         {-3, 0, -3},
         {-3, -3, -3} };
 short kirsch_mask_1[3][3] = {
         {-3, 5, 5},
         {-3, 0, 5},
         {-3, -3, -3} };
 short kirsch_mask_2[3][3] = {
         {-3, -3, 5},
         {-3, 0, 5},
         {-3, -3, 5} };
 short kirsch_mask_3[3][3] = {
         {-3, -3, -3},
         {-3, 0, 5},
         {-3, 5, 5} };
 short kirsch_mask_4[3][3] = {
         {-3, -3, -3},
         {-3, 0, -3},
         { 5, 5, 5} };
 short kirsch_mask_5[3][3] = {
         {-3, -3, -3},
         { 5, 0, -3},
         { 5, 5, -3} };
 short kirsch_mask_6[3][3] = {
```

```
        { 5, -3, -3},
        { 5, 0, -3},
        { 5, -3, -3} };
short kirsch_mask_7[3][3] = {
        { 5, 5, -3},
        { 5, 0, -3},
        {-3, -3, -3} };

  /* masks for prewitt operator */
short prewitt_mask_0[3][3] = {
        { 1, 1, 1},
        { 1, -2, 1},
        {-1, -1, -1} };
short prewitt_mask_1[3][3] = {
        { 1, 1, 1},
        { 1, -2, -1},
        { 1, -1, -1} };
short prewitt_mask_2[3][3] = {
        { 1, 1, -1},
        { 1, -2, -1},
        { 1, 1, -1} };
short prewitt_mask_3[3][3] = {
        { 1, -1, -1},
        { 1, -2, -1},
        { 1, 1, 1} };
short prewitt_mask_4[3][3] = {
        {-1, -1, -1},
        { 1, -2, 1},
        { 1, 1, 1} };
short prewitt_mask_5[3][3] = {
        {-1, -1, 1},
        {-1, -2, 1},
        { 1, 1, 1} };
short prewitt_mask_6[3][3] = {
        {-1, 1, 1},
        {-1, -2, 1},
        {-1, 1, 1} };
short prewitt_mask_7[3][3] = {
        { 1, 1, 1},
        {-1, -2, 1},
        {-1, -1, 1} };
 /* masks for sobel operator */
short sobel_mask_0[3][3] = {
        { 1, 2, 1},
```

```
          { 0, 0, 0},
          {-1, -2, -1} };
short sobel_mask_1[3][3] = {
          { 2, 1, 0},
          { 1, 0, -1},
          { 0, -1, -2} };
short sobel_mask_2[3][3] = {
          { 1, 0, -1},
          { 2, 0, -2},
          { 1, 0, -1} };
short sobel_mask_3[3][3] = {
          { 0, -1, -2},
          { 1, 0, -1},
          { 2, 1, 0} };
short sobel_mask_4[3][3] = {
          {-1, -2, -1},
          { 0, 0, 0},
          { 1, 2, 1} };
short sobel_mask_5[3][3] = {
          {-2, -1, 0},
          {-1, 0, 1},
          { 0, 1, 2} };
short sobel_mask_6[3][3] = {
          {-1, 0, 1},
          {-2, 0, 2},
          {-1, 0, 1} };
short sobel_mask_7[3][3] = {
          { 0, 1, 2},
          {-1, 0, 1},
          {-2, -1, 0} };
/**************************************************
```

detect_edges function detects edges in an area of one image and sends the result to another image on disk. It reads the input image from disk, calls a convolution function, and then writes the result out to disk. If needed, it allocates space on disk for the output image.

```
          **************************************************/
detect_edges(the_image, out_image,
       detect_type, threshold, high,
       rows, cols, bits_per_pixel)
   int detect_type, high, threshold;
   long rows, cols, bits_per_pixel;
   short **the_image, **out_image;
{
  perform_convolution(the_image, out_image,
```

```
                detect_type, threshold,
                rows, cols,
                bits_per_pixel,
                high);
   fix_edges(out_image, 1, rows, cols);
} /* ends detect_edges */

      /********************************************************
perform_convolution function performs convolution between the input image and
eight 3x3 masks. The result is placed in the out_image.
      ********************************************************/
perform_convolution(image, out_image,
                detect_type, threshold,
                rows, cols, bits_per_pixel, high)
      short **image,
          **out_image;
      int detect_type, high, threshold;
      long rows, cols, bits_per_pixel;
{
      char response[80];
      int a,
          b,
          i,
          is_present,
          j,
          sum;
   short mask_0[3][3],
          mask_1[3][3],
          mask_2[3][3],
          mask_3[3][3],
          mask_4[3][3],
          mask_5[3][3],
          mask_6[3][3],
          mask_7[3][3],
          max,
          min,
          new_hi,
          new_low;

   setup_masks(detect_type, mask_0, mask_1,
          mask_2, mask_3, mask_4, mask_5,
          mask_6, mask_7);
          new_hi = 250;
          new_low = 16;
          if(bits_per_pixel == 4){
```

```
              new_hi = 10;
              new_low = 3;
  }
  min = 0;
  max = 255;
  if(bits_per_pixel == 4)
     max = 16;
     /* clear output image array */
  for(i=0; i<rows; i++)
     for(j=0; j<cols; j++)
        out_image[i][j] = 0;
  printf("\n");

     for(i=1; i<rows-1; i++){
  if( (i%10) == 0){ printf("%4d", i); }
     for(j=1; j<cols-1; j++){

           /* Convolve for all 8 directions */
           /* 0 direction */
        sum = 0;
        for(a=-1; a<2; a++){
           for(b=-1; b<2; b++){
              sum = sum + image[i+a][j+b] *
                 mask_0[a+1][b+1];
           }
        }
           if(sum > max) sum = max;
           if(sum < 0) sum = 0;
        if(sum > out_image[i][j])
           out_image[i][j] = sum;

           /* 1 direction */
        sum = 0;
        for(a=-1; a<2; a++){
           for(b=-1; b<2; b++){
              sum = sum + image[i+a][j+b] * mask_1[a+1][b+1];
           }
        }
           if(sum > max) sum = max;
           if(sum < 0) sum = 0;
              /* Correction 12-27-92
              see file header for
              details. */
        if(sum > out_image[i][j])
           out_image[i][j] = sum;
```

```
    /* 2 direction */
sum = 0;
for(a=-1; a<2; a++){
    for(b=-1; b<2; b++){
        sum = sum + image[i+a][j+b] * mask_2[a+1][b+1];
    }
}
    if(sum > max) sum = max;
    if(sum < 0) sum = 0;
        /* Correction 12-27-92
            see file header for
            details. */
    if(sum > out_image[i][j])
        out_image[i][j] = sum;

        /* 3 direction */
sum = 0;
for(a=-1; a<2; a++){
    for(b=-1; b<2; b++){
        sum = sum + image[i+a][j+b] * mask_3[a+1][b+1];
    }
}
    if(sum > max) sum = max;
    if(sum < 0) sum = 0;
        /* Correction 12-27-92
            see file header for
            details. */
    if(sum > out_image[i][j])
        out_image[i][j] = sum;
        /* 4 direction */
    sum = 0;
    for(a=-1; a<2; a++){
        for(b=-1; b<2; b++){
            sum = sum + image[i+a][j+b] * mask_4[a+1][b+1];
        }
    }
    if(sum > max) sum = max;
    if(sum < 0) sum = 0;
        /* Correction 12-27-92
            see file header for
            details. */
    if(sum > out_image[i][j])
        out_image[i][j] = sum;
        /* 5 direction */
```

```
           sum = 0;
           for(a=-1; a<2; a++){
              for(b=-1; b<2; b++){
                 sum = sum + image[i+a][j+b] * mask_5[a+1][b+1];
              }
           }
              if(sum > max) sum = max;
              if(sum < 0) sum = 0;
                 /* Correction 12-27-92
                    see file header for
                    details. */
              if(sum > out_image[i][j])
                 out_image[i][j] = sum;
                 /* 6 direction */
              sum = 0;
              for(a=-1; a<2; a++){
                 for(b=-1; b<2; b++){
                    sum = sum + image[i+a][j+b] * mask_6[a+1][b+1];
           }
        }
     if(sum > max) sum = max;
     if(sum < 0) sum = 0;
        /* Correction 12-27-92
           see file header for
           details. */
     if(sum > out_image[i][j])
        out_image[i][j] = sum;
        /* 7 direction */
     sum = 0;
     for(a=-1; a<2; a++){
        for(b=-1; b<2; b++){
           sum = sum + image[i+a][j+b] * mask_7[a+1][b+1];
        }
     }
           if(sum > max) sum = max;
           if(sum < 0) sum = 0;
              /* Correction 12-27-92
                 see file header for
                 details. */
           if(sum > out_image[i][j])
              out_image[i][j] = sum;
           } /* ends loop over j */
        } /* ends loop over i */
```

```
    /* if desired, threshold the output image */
  if(threshold == 1){
     for(i=0; i<rows; i++){
        for(j=0; j<cols; j++){
           if(out_image[i][j] > high){
              out_image[i][j] = new_hi;
           }
           else{
              out_image[i][j] = new_low;
           }
        }
     }
  } /* ends if threshold == 1 */

} /* ends perform_convolution */

    /******************************************
    quick_edge function finds edges by using a single 3x3 mask.
    ******************************************/

quick_edge(the_image, out_image,
         threshold, high, rows, cols, bits_per_pixel)
      int high, threshold;
      long rows, cols, bits_per_pixel;
      short **the_image, **out_image;
{
    short a, b, i, j, k,
       length, max, new_hi, new_low,
       sum, width;
       new_hi = 250;
       new_low = 16;
       if(bits_per_pixel == 4){
          new_hi = 10;
          new_low = 3;
       }
       max = 255;
       if(bits_per_pixel == 4)
          max = 16;
             /* Do convolution over image array */
          printf("\n");
          for(i=1; i<rows-1; i++){
             if( (i%10) == 0) printf("%d", i);
             for(j=1; j<cols-1; j++){
                sum = 0;
                for(a=-1; a<2; a++){
```

```
               for(b=-1; b<2; b++){
                  sum = sum +
                     the_image[i+a][j+b] *
                     quick_mask[a+1][b+1];
                  }
               }
               if(sum < 0) sum = 0;
               if(sum > max) sum = max;
               out_image[i][j] = sum;

         } /* ends loop over j */
      } /* ends loop over i */
         /* if desired, threshold the output image */
      if(threshold == 1){
         for(i=0; i<rows; i++){
            for(j=0; j<cols; j++){
               if(out_image[i][j] > high){
                  out_image[i][j] = new_hi;
               }
               else{
                  out_image[i][j] = new_low;
               }
            }
         }
      } /* ends if threshold == 1 */

      fix_edges(out_image, 1,
            rows-1, cols-1);

   } /* ends quick_edge */

/************************************************
setup_masks function copies the mask values defined at the top of this file into the
mask arrays mask_0 through mask_7.
   ***********************************************/

setup_masks(detect_type, mask_0, mask_1, mask_2, mask_3,
   mask_4, mask_5, mask_6, mask_7)
   int detect_type;
   short mask_0[3][3],
         mask_1[3][3],
         mask_2[3][3],
         mask_3[3][3],
         mask_4[3][3],
         mask_5[3][3],
         mask_6[3][3],
```

```
            mask_7[3][3];
{
    int i, j;

    if(detect_type == KIRSCH){
        for(i=0; i<3; i++){
            for(j=0; j<3; j++){
                mask_0[i][j] = kirsch_mask_0[i][j];
                mask_1[i][j] = kirsch_mask_1[i][j];
                mask_2[i][j] = kirsch_mask_2[i][j];
                mask_3[i][j] = kirsch_mask_3[i][j];
                mask_4[i][j] = kirsch_mask_4[i][j];
                mask_5[i][j] = kirsch_mask_5[i][j];
                mask_6[i][j] = kirsch_mask_6[i][j];
                mask_7[i][j] = kirsch_mask_7[i][j];
            }
        }
}  /* ends if detect_type == KIRSCH */

if(detect_type == PREWITT){
        for(i=0; i<3; i++){
            for(j=0; j<3; j++){
                mask_0[i][j] = prewitt_mask_0[i][j];
                mask_1[i][j] = prewitt_mask_1[i][j];
                mask_2[i][j] = prewitt_mask_2[i][j];
                mask_3[i][j] = prewitt_mask_3[i][j];
                mask_4[i][j] = prewitt_mask_4[i][j];
                mask_5[i][j] = prewitt_mask_5[i][j];
                mask_6[i][j] = prewitt_mask_6[i][j];
                mask_7[i][j] = prewitt_mask_7[i][j];
            }
        }
    }  /* ends if detect_type == PREWITT */
    if(detect_type == SOBEL){
        for(i=0; i<3; i++){
            for(j=0; j<3; j++){
                mask_0[i][j] = sobel_mask_0[i][j];
                mask_1[i][j] = sobel_mask_1[i][j];
                mask_2[i][j] = sobel_mask_2[i][j];
                mask_3[i][j] = sobel_mask_3[i][j];
                mask_4[i][j] = sobel_mask_4[i][j];
                mask_5[i][j] = sobel_mask_5[i][j];
                mask_6[i][j] = sobel_mask_6[i][j];
                mask_7[i][j] = sobel_mask_7[i][j];
            }
```

```
        }
    } /* ends if detect_type == SOBEL */

} /* ends setup_masks */

#ifdef NEVER

    /*************************************************
 get_edge_options function queries the user for the parameters needed in order to
perform edge detection.
    *************************************************/
get_edge_options(detect_type, threshold, high, size)
    int *detect_type, *high, *size, *threshold;
{
    int not_finished, response;
    not_finished = 1;
    while(not_finished){

    printf("\nThe Edge Detector options are:\n");
    printf("\n\t1.  Type of edge detector is %d", *detect_type);
    printf("\n\t    (recall 1=Prewitt 2=Kirsch");
    printf("\n\t               3=Sobel          4=quick");
    printf("\n\t               5=homogeneity  6=difference");
    printf("\n\t               7=contrast      8=gaussian");
    printf("\n\t               10=range        11=variance");
    printf("\n\t2. Threshold output is %d (0=off 1=on)", *threshold);
    printf("\n\t3. High threshold is %d", *high);
    printf("\n\t4. Size is %d (gaussian only)", *size);
    printf("\n\nEnter choice (0 = no change) _\b");
    get_integer(&response);
    if(response == 0){
        not_finished = 0;
    }
    if(response == 1){
      printf("\n\nEnter type of edge detector");
      printf("\n\t    (recall 1=Prewitt          2=Kirsch");
      printf("\n\t    3=Sobel                    4=quick");
      printf("\n\t    5=homogeneity              6=difference");
      printf("\n\t    7=contrast                 8=gaussian");
      printf("\n\t    10=range                   11=variance");
      printf("\n _\b");
      get_integer(detect_type);
    }
    if(response == 2){
      printf("\n\nEnter threshold output (0=off 1=on)");
      printf("\n _\b");
```

```
      get_integer(threshold);
    }
    if(response == 3){
      printf("\n\nEnter high threshold");
      printf("\n _\b");
      get_integer(high);
    }
    if(response == 4){
      printf("\n\nEnter size for gaussian (7 or 9)");
      printf("\n _\b");
      get_integer(size);
     }
  } /* ends while not_finished */

} /* ends get_edge_options */
#endif
short e_mask[3][3] = {
        {-9, 0, -9},
        { 0, 36, 0},
        {-9, 0, -9} };
short contrast[3][3] = {
     { 1, 1, 1},
     { 1, 1, 1},
     { 1, 1, 1}};
```

```
      /**************************************************
```
  homogeneity function performs edge detection by looking for the absence of
an edge. The center of a 3x3 area is replaced by the absolute value of the max.
difference between the center point and its 8 neighbors.
```
      **************************************************/
```

```
homogeneity(the_image, out_image,
          rows, cols, bits_per_pixel,
          threshold, high)
   int   high, threshold;
   short  **the_image, **out_image;
   long   rows, cols, bits_per_pixel;
{
   int a, b, absdiff, absmax, diff, i, j,
      length, max, max_diff, new_hi, new_low, width;
   new_hi = 250;
   new_low = 16;
   if(bits_per_pixel == 4){
     new_hi = 10;
     new_low = 3;
```

```
        }
        max = 255;
        if(bits_per_pixel == 4)
            max = 16;
        for(i=0; i<rows; i++){
            for(j=0; j<cols; j++){
                out_image[i][j] = 0;
            }
        }
        for(i=1; i<rows-1; i++){
            if( (i%10) == 0) printf("%4d", i);
            for(j=1; j<cols-1; j++){
                max_diff = 0;
                for(a=-1; a<=1; a++){
                    for(b=-1; b<=1; b++){
                        diff = the_image[i][j] -
                            the_image[i+a][j+b];
                        absdiff = abs(diff);
                        if(absdiff > max_diff)
                            max_diff = absdiff;
                    } /* ends loop over b */
                } /* ends loop over a */
                out_image[i][j] = max_diff;
            } /* ends loop over j */
} /* ends loop over i */

    /* if desired, threshold the output image */
    if(threshold == 1){
        for(i=0; i<rows; i++){
            for(j=0; j<cols; j++){
                if(out_image[i][j] > high){
                    out_image[i][j] = new_hi;
                }
                else{
                    out_image[i][j] = new_low;
                }
            }
        }
    } /* ends if threshold == 1 */
} /* ends homogeneity */

    /***********************************************
```

difference_edge function performs edge detection by looking at the differences in the pixels that surround the center point of a 3x3 area. It replaces the center point with the absolute value of the max difference of:

```
        upper left - lower right
        upper right - lower left
        left - right
        top - bottom
    ***************************************************/

difference_edge(the_image, out_image,
        rows, cols, bits_per_pixel,
        threshold, high)
    int high, threshold;
    short **the_image, **out_image;
    long rows, cols, bits_per_pixel;
{
    int a, b, absdiff, absmax, diff, i, j,
        length, max, max_diff, new_hi, new_low, width;
    new_hi = 250;
    new_low = 16;
    if(bits_per_pixel == 4){
        new_hi = 10;
        new_low = 3;
    }
    max = 255;
    if(bits_per_pixel == 4)
        max = 16;
    for(i=0; i<rows; i++)
        for(j=0; j<cols; j++)
            out_image[i][j] = 0;
    for(i=1; i<rows-1; i++){
if( (i%10) == 0) printf("%4d", i);
for(j=1; j<cols-1; j++){
    max_diff = 0;
    absdiff = abs(the_image[i-1][j-1] -
            the_image[i+1][j+1]);
    if(absdiff > max_diff) max_diff = absdiff;

    absdiff = abs(the_image[i-1][j+1] -
            the_image[i+1][j-1]);
    if(absdiff > max_diff) max_diff = absdiff;

    absdiff = abs(the_image[i][j-1] -
            the_image[i][j+1]);
    if(absdiff > max_diff) max_diff = absdiff;

    absdiff = abs(the_image[i-1][j] -
            the_image[i+1][j]);
    if(absdiff > max_diff) max_diff = absdiff;
```

```
        out_image[i][j] = max_diff;
    } /* ends loop over j */
  } /* ends loop over i */

      /* if desired, threshold the output image */
  if(threshold == 1){
      for(i=0; i<rows; i++){
          for(j=0; j<cols; j++){
              if(out_image[i][j] > high){
                  out_image[i][j] = new_hi;
          }
        else{
          out_image[i][j] = new_low;
          }
        }
      }
  } /* ends if threshold == 1 */

    } /* ends difference_edge */

    /**************************************************
```

contrast_edge edge detector uses the basic quick edge detector mask and then divides the result by a contrast smooth mask. This implements Johnson's contrast-based edge detector.

```
    **************************************************/

contrast_edge(the_image, out_image,
      rows, cols, bits_per_pixel,
      threshold, high)
  int high, threshold;
  short **the_image, **out_image;
  long rows, cols, bits_per_pixel;
{
  int ad, d;
  int a, b, absdiff, absmax, diff, i, j,
      length, max, new_hi, new_low,
      sum_d, sum_n, width;
  new_hi = 250;
  new_low = 16;
  if(bits_per_pixel == 4){
      new_hi = 10;
      new_low = 3;
  }
  max = 255;
  if(bits_per_pixel == 4)
      max = 16;
```

```
    for(i=0; i<rows; i++)
        for(j=0; j<cols; j++)
            out_image[i][j] = 0;
    for(i=1; i<rows-1; i++){
      if( (i%10) == 0) printf("%4d", i);
      for(j=1; j<cols-1; j++){
        sum_n = 0;
        sum_d = 0;
        for(a=-1; a<2; a++){
            for(b=-1; b<2; b++){
                sum_n = sum_n + the_image[i+a][j+b] *
                    e_mask[a+1][b+1];
                sum_d = sum_d + the_image[i+a][j+b] *
                    contrast[a+1][b+1];
            }
        }
        d = sum_d / 9;
        if(d == 0)
            d = 1;
        out_image[i][j] = sum_n/d;
        if(out_image[i][j] > max)
            out_image[i][j] = max;
        if(out_image[i][j] < 0)
            out_image[i][j] = 0;

      } /* ends loop over j */
    } /* ends loop over i */
    /* if desired, threshold the output image */
    if(threshold == 1){
        for(i=0; i<rows; i++){
            for(j=0; j<cols; j++){
                if(out_image[i][j] > high){
                    out_image[i][j] = new_hi;
                }
                else{
                    out_image[i][j] = new_low;
                }
            }
        }
    } /* ends if threshold == 1 */
} /* ends contrast_edge */

    /*******************************************
```
range edge detector performs the range operation. It replaces the pixel at the center
of a 3x3, 5x5, etc. area with the max–min for that area.
```
    *******************************************/
```

```
range(the_image, out_image,
      rows, cols, bits_per_pixel,
      size, threshold, high)
   int high, threshold, size;
   short **the_image,
         **out_image;
   long rows, cols, bits_per_pixel;
{
   int a, b, count, i, j, k,
      new_hi, new_low, length,
      sd2, sd2p1, ss, width;
short *elements;

sd2 = size/2;
sd2p1 = sd2 + 1;
   /************************************************
Allocate the elements array large enough to hold size*size shorts.
   ************************************************/

ss      = size*size;
elements = (short *) malloc(ss * sizeof(short));

new_hi = 250;
new_low = 16;
if(bits_per_pixel == 4){
   new_hi = 10;
   new_low = 3;
}

   /**************************
* Loop over image array
   **************************/

printf("\n");
for(i=sd2; i<rows-sd2; i++){
   if( (i%10) == 0) printf("%4d", i);
   for(j=sd2; j<cols-sd2; j++){
      count = 0;
      for(a=-sd2; a<sd2p1; a++){
         for(b=-sd2; b<sd2p1; b++){
            elements[count] = the_image[i+a][j+b];
            count++;
         }
      }
      sort_elements(elements, &ss);
      out_image[i][j] = elements[ss-1]-elements[0];
```

Appendices 273

```
      } /* ends loop over j */
    } /* ends loop over i */

    /* if desired, threshold the output image */
   if(threshold == 1){
      for(i=0; i<rows; i++){
         for(j=0; j<cols; j++){
            if(out_image[i][j] > high){
               out_image[i][j] = new_hi;
            }
            else{
               out_image[i][j] = new_low;
            }
         }
      }
   } /* ends if threshold == 1 */

 free(elements);

} /* ends range */

   /**************************************************
Variance function replaces the pixel in the center of a 3x3 area with the square
root of the sum of squares of the differences between the center pixel and its eight
neighbors.
   **************************************************/

variance(the_image, out_image,
      rows, cols, bits_per_pixel,
      threshold, high)
   int high, threshold;
   short **the_image,
      **out_image;
   long rows, cols, bits_per_pixel;
{
   int   a, b, i, j, length,
         max, new_hi, new_low, width;
   long diff;
   unsigned long sum, tmp;

   new_hi = 250;
   new_low = 16;
   if(bits_per_pixel == 4){
      new_hi = 10;
      new_low = 3;
   }
   max = 255;
```

```
if(bits_per_pixel == 4)
    max = 16;
for(i=1; i<rows-1; i++){
    if( (i%10) == 0) printf("%4d", i);
    for(j=1; j<cols-1; j++){
        sum = 0;
        for(a=-1; a<=1; a++){
            for(b=-1; b<=1; b++){
                if( a!=0 && b!=0){
                diff = 0;
                diff = the_image[i][j] -
                    the_image[i+a][j+b];
                tmp = diff*diff;
                sum = sum + tmp;
                }
            }
        }
}
if(sum < 0)
    printf("\nWHAT? sum < 0, %ld, diff=%d", sum, diff);
    sum = sqrt(sum);
    if(sum > max) sum = max;
    out_image[i][j] = sum;
    } /* ends loop over j */
} /* ends loop over i */
  /* if desired, threshold the output image */
if(threshold == 1){
    for(i=0; i<rows; i++){
        for(j=0; j<cols; j++){
            if(out_image[i][j] > high){
                out_image[i][j] = new_hi;
            }
            else{
                out_image[i][j] = new_low;
            }
        }
    }
    } /* ends if threshold == 1 */
} /* ends variance */

short enhance_mask[3][3] = {
        {-1, 0, -1},
        { 0, 4, 0},
        {-1, 0, -1} };
short g7[7][7] = {
```

```
      { 0, 0, -1, -1, -1, 0, 0},
      { 0, -2, -3, -3, -3, -2, 0},
      { -1, -3, 5, 5, 5, -3, -1},
      { -1, -3, 5, 16, 5, -3, -1},
      { -1, -3, 5, 5, 5, -3, -1},
      { 0, -2, -3, -3, -3, -2, 0},
      { 0, 0, -1, -1, -1, 0, 0}};
short g9[9][9] = {
      { 0, 0, 0, -1, -1, -1, 0, 0, 0},
      { 0, -2, -3, -3, -3, -3, -3, -2, 0},
      { 0, -3, -2, -1, -1, -1, -2, -3, 0},
      { -1, -3, -1, 9, 9, 9, -1, -3, -1},
      { -1, -3, -1, 9, 19, 9, -1, -3, -1},
      { -1, -3, -1, 9, 9, 9, -1, -3, -1},
      { 0, -3, -2, -1, -1, -1, -2, -3, 0},
      { 0, -2, -3, -3, -3, -3, -3, -2, 0},
      { 0, 0, 0, -1, -1, -1, 0, 0, 0}};

/************************************************
 * gaussian_edge(...
 ************************************************/
gaussian_edge(the_image, out_image,
       rows, cols, bits_per_pixel,
       size, threshold, high)
    int high, size, threshold;
    short **the_image,
       **out_image;
    long rows, cols, bits_per_pixel;
{
    char response[80];
    long sum;
    int a, b, absdiff, absmax, diff, i, j,
       length, lower, max, new_hi, new_low,
       scale, starti, stopi, startj, stopj,
       upper, width;

    new_hi = 250;
    new_low = 16;
    if(bits_per_pixel == 4){
       new_hi = 10;
       new_low = 3;
    }
    max = 255;
    if(bits_per_pixel == 4)
       max = 16;
```

```
if(size == 7){
    lower = -3;
    upper = 4;
    starti = 3;
    startj = 3;
    stopi = rows-3;
    stopj = cols-3;
    scale = 2;
}
if(size == 9){
    lower = -4;
    upper = 5;
    starti = 4;
    startj = 4;
    stopi = rows-4;
    stopj = cols-4;
    scale = 2;
}
for(i=0; i<rows; i++)
    for(j=0; j<cols; j++)
        out_image[i][j] = 0;
for(i=starti; i<stopi; i++){
    if ( (i%10) == 0) printf(" i=%d", i);
    for(j=startj; j<stopj; j++){
    sum = 0;
    for(a=lower; a<upper; a++){
        for(b=lower; b<upper; b++){
            if(size == 7)
                sum = sum + the_image[i+a][j+b] *
                    g7[a+3][b+3];
    if(size == 9)
        sum = sum + the_image[i+a][j+b] *
            g9[a+4][b+4];
  } /* ends loop over a */
} /* ends loop over b */

if(sum < 0) sum = 0;
if(sum > max) sum = max;
out_image[i][j] = sum;

    } /* ends loop over j */
} /* ends loop over i */

  /* if desired, threshold the output image */
if(threshold == 1){
```

```
    for(i=0; i<rows; i++){
        for(j=0; j<cols; j++){
            if(out_image[i][j] > high){
                out_image[i][j] = new_hi;
            }
            else{
                out_image[i][j] = new_low;
            }
        }
    }
  } /* ends if threshold == 1 */

} /* ends gaussian_edge */
```

```
    /*******************************************
```

enhance_edges function enhances the edges in an input image and writes the enhanced result to an output image. It operates much the same way as detect_edges except it uses only one type of mask. The threshold and high parameters perform a different role in this function. The threshold parameter does not exist. The high parameter determines if the edge is strong enough to enhance or change the input image.

```
    *******************************************/
```

```
enhance_edges(the_image, out_image,
        rows, cols, bits_per_pixel, high)
    int high;
    short **the_image,
            **out_image;
    long rows, cols, bits_per_pixel;
  {
    int a, b, i, j, k,
        length, max, new_hi,
        new_lo, sum, width;
    max = 255;
    if(bits_per_pixel == 4)
        max = 16;
            /* Do convolution over image array */
    for(i=1; i<rows-1; i++){
        if( (i%10) == 0) printf("%d", i);
        for(j=1; j<cols-1; j++){
            sum = 0;
            for(a=-1; a<2; a++){
                for(b=-1; b<2; b++){
                    sum = sum +
                        the_image[i+a][j+b] *
```

```
                        enhance_mask[a+1][b+1];
            }
                }
            if(sum < 0) sum = 0;
            if(sum > max) sum = max;
            if(sum > high)
                out_image[i][j] = max;
            else
                out_image[i][j] = the_image[i][j];
        } /* ends loop over j */
    } /* ends loop over i */
} /* ends enhance_edges */
```

B.11 Skeleton

```
/*********************************************
```

Functions in this file dilate objects and perform opening and closing without removing or joining objects.

```
    /*********************************************
```

special_opening: Opening is erosion followed by dilation. This routine will use the thinning erosion routine. This will not allow an object to erode to nothing. The number parameter specifies how many erosions to perform before doing one dilation.

```
    *******************************************/

special_opening(the_image, out_image,
        value, threshold, number,
        rows, cols)
    int number;
    short **the_image,
        **out_image,
        threshold, value;
    long cols, rows;
{
    int a, b, count, i, j, k;

thinning(the_image, out_image,
      value, threshold, 1,
      rows, cols);
if(number > 1){
    count = 1;
    while(count < number){
        count++;
        thinning(the_image, out_image,
          value, threshold, 1,
```

```
        rows, cols);
      } /* ends while */
  } /* ends if number > 1 */

  dilation(the_image, out_image,
        value, threshold,
        rows, cols);
} /* ends special_opening */

thinning(the_image, out_image,
      value, threshold, once_only,
      rows, cols)
   int once_only;
   short **the_image,
      **out_image,
      threshold, value;
   long cols, rows;
{
   int a, b, big_count, count, i, j, k,
      not_finished;

   for(i=0; i<rows; i++)
      for(j=0; j<cols; j++)
         out_image[i][j] = the_image[i][j];

   not_finished = 1;
   while(not_finished){

      if(once_only == 1)
         not_finished = 0;
      big_count = 0;

   /*************************
Scan left to right, Look for 0-value transition
      *************************/

      printf("\n");
      for(i=1; i<rows-1; i++){
         if( (i%10) == 0) printf("%3d", i);
         for(j=1; j<cols-1; j++){
            if(the_image[i][j-1] == 0 &&
               the_image[i][j] == value){
               count = 0;
               for(a=-1; a<=1; a++){
                  for(b=-1; b<=1; b++){
                     if(the_image[i+a][j+b] == 0)
                        count++;
```

```
            } /* ends loop over b */
         } /* ends loop over a */
         if(count > threshold){
            if(can_thin(the_image, i, j, value)){
               out_image[i][j] = 0;
               big_count++;
            } /* ends if can_thin */
         } /* ends if count > threshold */
      } /* ends if the_image == value */
   } /* ends loop over j */
} /* ends loop over i */

   /***********************************
                  Copy the output back to the input.
   ***********************************/

for(i=0; i<rows; i++)
     for(j=0; j<cols; j++)
        the_image[i][j] = out_image[i][j];

printf("\n");
for(i=1; i<rows-1; i++){
   if( (i%10) == 0) printf("%3d", i);
   for(j=1; j<cols-1; j++){
      if(the_image[i][j+1] == 0 &&
         the_image[i][j] == value){
         count = 0;
         for(a=-1; a<=1; a++){
            for(b=-1; b<=1; b++){
               if(the_image[i+a][j+b] == 0)
                  count++;
            } /* ends loop over b */
         } /* ends loop over a */
         if(count > threshold){
            if(can_thin(the_image, i, j, value)){
               out_image[i][j] = 0;
               big_count++;
            } /* ends if can_thin */
         } /* ends if count > threshold */
      } /* ends if the_image == value */
   } /* ends loop over j */
} /* ends loop over i */

   /***********************************
   * Copy the output back to the input.
   ***********************************/
```

```
for(i=0; i<rows; i++)
   for(j=0; j<cols; j++)
      the_image[i][j] = out_image[i][j];

   /**************************
            Scan top to bottom, Look for 0-value transition
*
      **************************/

printf("\n");
for(j=1; j<cols-1; j++){
   if( (j%10) == 0) printf("%3d", j);
   for(i=1; i<rows-1; i++){
         if(the_image[i-1][j] == 0 &&
            the_image[i][j] == value){
         count = 0;
         for(a=-1; a<=1; a++){
            for(b=-1; b<=1; b++){
               if(the_image[i+a][j+b] == 0)
                  count++;
            } /* ends loop over b */
         } /* ends loop over a */
         if(count > threshold){
            if(can_thin(the_image, i, j, value)){
               out_image[i][j] = 0;
               big_count++;
            } /* ends if can_thin */
         } /* ends if count > threshold */
      } /* ends if the_image == value */
   } /* ends loop over i */
} /* ends loop over j */

   /*************************************
            Copy the output back to the input.
      *************************************/

for(i=0; i<rows; i++)
   for(j=0; j<cols; j++)
         the_image[i][j] = out_image[i][j];

   /**************************
      Scan bottom to top
      **************************/

printf("\n");
for(j=1; j<cols-1; j++){
```

```
    if( (j%10) == 0) printf("%3d", j);
    for(i=1; i<rows-1; i++){
        if(the_image[i+1][j] == 0 &&
            the_image[i][j] == value){
            count = 0;
            for(a=-1; a<=1; a++){
                for(b=-1; b<=1; b++){
                    if(the_image[i+a][j+b] == 0)
                        count++;
                } /* ends loop over b */
            } /* ends loop over a */
            if(count > threshold){
                if(can_thin(the_image, i, j, value)){
                    out_image[i][j] = 0;
                    big_count++;
                } /* ends if can_thin */
            } /* ends if count > threshold */
        } /* ends if the_image == value */
    } /* ends loop over i */
} /* ends loop over j */

    /**************************************
            Copy the output back to the input.
    **************************************/

for(i=0; i<rows; i++)
    for(j=0; j<cols; j++)
        the_image[i][j] = out_image[i][j];

    printf("\n\nThinned %d pixels", big_count);
if(big_count == 0)
    not_finished = 0;
else{
    for(i=0; i<rows; i++)
        for(j=0; j<cols; j++)
            the_image[i][j] = out_image[i][j];
    } /* ends else */

  } /* ends while not_finished */

  /****
  fix_edges(out_image, 3, rows, cols);
  *****/

} /* ends thinning */
```

```
/*****************************************
can_thin : Look at the neighbors of the center pixel.
 * If a neighbor == value, then it must have a neighbor == value other than the
center pixel.
 *****************************************/

can_thin(the_image, i, j, value)
   int i, j;
   short **the_image, value;
{
   int a, b, c, d, count,
       no_neighbor, one=1, zero=0;
   short temp[3][3];

   /*****************************************
Copy the center pixel and its neighbors to the temp array.
 *****************************************/

   for(a=-1; a<2; a++)
      for(b=-1; b<2; b++)
         temp[a+1][b+1] = the_image[i+a][b+j];
   temp[1][1] = 0;

   /*****************************************
             Check the non-zero pixels in temp.
 *****************************************/

   for(a=0; a<3; a++){
      for(b=0; b<3; b++){
         if(temp[a][b] == value){
            temp[a][b] = 0;

   /********************************
Check the neighbors of this pixel, if there is a single non-zero neighbor, set
no_neighbor = 0.
 ********************************/

            no_neighbor = 1;
            for(c=-1; c<2; c++){
               for(d=-1; d<2; d++){
                  if( ((a+c) >= 0) &&
                     ((a+c) <= 2) &&
                     ((b+d) >= 0) &&
                     ((b+d) <= 2)){
                     if(temp[a+c][b+d] == value){
                        no_neighbor = 0;
                     } /* ends if temp == value */
```

```
        } /* ends if part of temp array */
    } /* ends loop over d */
} /* ends loop over c */
temp[a][b] = value;

    /********************************
```

If the non-zero pixel did not have any non-zero neighbors, no_neighbor still equals
1 and we cannot thin, therefore, return zero.
```
    *********************************/

   if(no_neighbor){
     return(zero);
     }
   } /* ends if temp[a][b] == value */
  } /* ends loop over b */
 } /* ends loop over a */

   return(one);
} /* ends can_thin */

   /*******************************************
```
special_closing: Closing is dilation followed by erosion.
```
   *******************************************/

special_closing(the_image, out_image,
        value, threshold, number,
        rows, cols)
   int number;
   short **the_image,
       **out_image,
       threshold, value;
    long cols, rows;
{
   int a, b, count, i, j, k;

   dilate_not_join(the_image, out_image,
           value, threshold,
           rows, cols);
   if(number > 1){
     count = 1;
     while(count < number){
         count++;
         dilate_not_join(the_image, out_image,
                 value, threshold,
                 rows, cols);
     } /* ends while */
   } /* ends if number > 1 */
```

```
      erosion(the_image, out_image,
          value, threshold,
          rows, cols);
   } /* ends special_closing */
dilate_not_join(the_image, out_image,
          value, threshold,
          rows, cols)
   short **the_image,
          **out_image,
          threshold, value;
   long cols, rows;
{
   int a, b, count, i, j, k;
   for(i=0; i<rows; i++)
      for(j=0; j<cols; j++)
         out_image[i][j] = the_image[i][j];
       printf("\n");
       for(i=1; i<rows-1; i++){
          if( (i%10) == 0) printf("%3d", i);
          for(j=1; j<cols-1; j++){
             if(the_image[i][j-1] == value &&
                the_image[i][j] == 0){
                count = 0;
                for(a=-1; a<=1; a++){
                   for(b=-1; b<=1; b++){
                      if(the_image[i+a][j+b]==value)
                         count++;
                   } /* ends loop over b */
                } /* ends loop over a */
                if(count > threshold){
                   if(can_dilate(the_image,i,j,value)){
                      out_image[i][j] = value;
                   } /* ends if can_dilate */
                } /* ends if count > threshold */
             } /* ends if the_image == value */
          } /* ends loop over j */
       } /* ends loop over i */
   for(i=0; i<rows; i++)
      for(j=0; j<cols; j++)
         the_image[i][j] = out_image[i][j];
   printf("\n");
   for(i=1; i<rows-1; i++){
      if( (i%10) == 0) printf("%3d", i);
      for(j=1; j<cols-1; j++){
```

```
            if(the_image[i][j+1] == value &&
                the_image[i][j] == 0){
                count = 0;
                for(a=-1; a<=1; a++){
                    for(b=-1; b<=1; b++){
                        if(the_image[i+a][j+b]==value)
                            count++;
                } /* ends loop over b */
        } /* ends loop over a */
        if(count > threshold){
                if(can_dilate(the_image,i,j,value)){
                    out_image[i][j] = value;
                } /* ends if can_dilate */
            } /* ends if count > threshold */
        } /* ends if the_image == value */
    } /* ends loop over j */
} /* ends loop over i */
for(i=0; i<rows; i++)
  for(j=0; j<cols; j++)
    the_image[i][j] = out_image[i][j];

printf("\n");
for(j=1; j<cols-1; j++){
  if( (j%10) == 0) printf("%3d", j);
  for(i=1; i<rows-1; i++){
    if(the_image[i-1][j] == value &&
        the_image[i][j] == 0){
        count = 0;
        for(a=-1; a<=1; a++){
            for(b=-1; b<=1; b++){
                if(the_image[i+a][j+b]==value)
                    count++;
        } /* ends loop over b */
    } /* ends loop over a */
    if(count > threshold){
        if(can_dilate(the_image,i,j,value)){
            out_image[i][j] = value;
        } /* ends if can_dilate */
    } /* ends if count > threshold */
    } /* ends if the_image == value */
  } /* ends loop over i */
} /* ends loop over j */
for(i=0; i<rows; i++)
  for(j=0; j<cols; j++)
```

```
        the_image[i][j] = out_image[i][j];
printf("\n");
for(j=1; j<cols-1; j++){
   if( (j%10) == 0) printf("%3d", j);
   for(i=1; i<rows-1; i++){
     if(the_image[i+1][j] == value &&
        the_image[i][j] == 0){
        count = 0;
        for(a=-1; a<=1; a++){
            for(b=-1; b<=1; b++){
                if(the_image[i+a][j+b]==value)
                    count++;
            } /* ends loop over b */
        } /* ends loop over a */
        if(count > threshold){
           if(can_dilate(the_image,i,j,value)){
               out_image[i][j] = value;
           } /* ends if can_dilate */
        } /* ends if count > threshold */
     } /* ends if the_image == value */
   } /* ends loop over i */
 } /* ends loop over j */
 for(i=0; i<rows; i++)
   for(j=0; j<cols; j++)
     the_image[i][j] = out_image[i][j];
 /****
 fix_edges(out_image, 3, rows, cols);
 *****/
} /* ends dilate_not_join */
```

```
    /*******************************************
can_dilate function decides if you can dilate (set to value) a pixel without joining
two separate objects in a 3x3 area.
    *******************************************/
```

```
can_dilate(the_image, i, j, value)
   int i, j;
   short **the_image, value;
{
   int a, b, c, d, count, found=0,
       no_neighbor,
       stack_pointer=-1,
       stack_empty=1,
       stack[12][2],
```

```
        pop_a, pop_b,
        one=1,
        zero=0;
    short first_value, label = 2, temp[3][3];
    for(a=-1; a<2; a++)
      for(b=-1; b<2; b++)
        temp[a+1][b+1] = the_image[i+a][b+j];
      for(a=0; a<3; a++){
        for(b=0; b<3; b++){
            stack_empty = 1;
            stack_pointer = -1;
            if(temp[a][b] == value){
                little_label_and_check(temp, stack, label,
&stack_empty,
&stack_pointer,
                            a, b, value);
              found = 1;
            } /* ends if temp == value */
            while(stack_empty == 0){
                pop_a = stack[stack_pointer][0]; /* POP */
                pop_b = stack[stack_pointer][1]; /* POP */
                --stack_pointer;
                if(stack_pointer <= 0){
                  stack_pointer = 0;
                  stack_empty = 1;
                } /* ends if stack_pointer */
                little_label_and_check(temp, stack, label,
&stack_empty,
&stack_pointer,
                            pop_a, pop_b, value);
          } /* ends while stack_empty == 0 */
          if(found){
            found = 0;
          label++;
        } /* ends if object_found */
      } /* ends loop over b */
  } /* ends loop over a */
  first_value = -1;
  for(a=0; a<3; a++){
    for(b=0; b<3; b++){
      if(temp[a][b] != 0 &&
        first_value == -1){
        first_value = temp[a][b];
      }
```

```
      if(temp[a][b] != 0 &&
         first_value != -1){
         if(temp[a][b] != first_value){
            return(zero);
      }
    }
  } /* ends loop over b */
 } /* ends loop over a */

 return(one);

} /* ends can_dilate */
```

```
     /*********************************************
little_label_and_check function labels the objects in a 3x3 area.
     *********************************************/
```

```
little_label_and_check(temp, stack, label, stack_empty,
              stack_pointer, a, b, value)
   int a, b, stack[12][2],
       *stack_empty, *stack_pointer;
   short temp[3][3], label, value;
{
   int c, d;

   temp[a][b] = label;
   for(c=a-1; c<=a+1; c++){
      for(d=b-1; d<=b+1; d++){
         if(c >= 0 &&
            c <= 2 &&
            d >= 0 &&
            d <= 2)
            if(temp[c][d] == value){ /* PUSH */
               *stack_pointer = *stack_pointer + 1;
               stack[*stack_pointer][0] = c;
               stack[*stack_pointer][1] = d;
               *stack_empty = 0;
            } /* ends if temp == value */
      } /* ends loop over d */
   } /* ends loop over c */
} /* ends little_label_and_check */
```

```
     /*********************************************
edm function calculates the Euclidean distance measure for objects in an image. It
calculates the distance from any pixel=value to the nearest zero pixel
     *********************************************/
```

```
edm(the_image, out_image,
    value, rows, cols)
  short **the_image,
    **out_image,
    value;
  long cols, rows;
{
  int a, b, count, i, j, k;

  for(i=0; i<rows; i++)
    for(j=0; j<cols; j++)
      out_image[i][j] = 0;
  printf("\n");

  for(i=0; i<rows; i++){
    if( (i%10) == 0) printf("%3d", i);
    for(j=0; j<cols; j++){
      if(the_image[i][j] == value)
        out_image[i][j] = distance_8(the_image,
                          i, j,
                          value,
                          rows, cols);
      } /* ends loop over j */
    } /* ends loop over i */

} /* ends edm */

      /******************************************
distance_8 function finds the distance from a pixel to the nearest zero pixel. It
search in all eight directions.
      ******************************************/

distance_8(the_image, a, b, value, rows, cols)
    int a, b;
    short **the_image, value;
    long cols, rows;
{
    int i, j, measuring;
    short dist1 = 0,
        dist2 = 0,
        dist3 = 0,
        dist4 = 0,
        dist5 = 0,
        dist6 = 0,
        dist7 = 0,
        dist8 = 0,
```

```
        result = 0;

     /* straight up */
measuring = 1;
i = a;
j = b;
while(measuring){
  i--;
  if(i >= 0){
     if(the_image[i][j] == value)
        dist1++;
     else
        measuring = 0;
  }
  else
     measuring = 0;
} /* ends while measuring */
result = dist1;

     /* straight down */
measuring = 1;
i = a;
j = b;
 while(measuring){
    i++;
    if(i <= rows-1){
       if(the_image[i][j] == value)
          dist2++;
       else
          measuring = 0;
    }
    else
       measuring = 0;
} /* ends while measuring */
if(dist2 <= result)
   result = dist2;

  /* straight left */
measuring = 1;
i = a;
j = b;
while(measuring){
   j--;
   if(j >= 0){
      if(the_image[i][j] == value)
```

```
         dist3++;
      else
          measuring = 0;
  }
  else
    measuring = 0;
} /* ends while measuring */
if(dist3 <= result)
  result = dist3;
   /* straight right */
measuring = 1;
i = a;
j = b;
while(measuring){
   j++;
  if(j <= cols-1){
   if(the_image[i][j] == value)
      dist4++;
   else
     measuring = 0;
   }
   else
     measuring = 0;
} /* ends while measuring */
if(dist4 <= result)
    result = dist4;
    /* left and up */
measuring = 1;
i = a;
j = b;
while(measuring){
    j--;
    i--;
    if(j > = 0 && i>=0){
       if(the_image[i][j] == value)
     dist5++;
  else
    measuring = 0;
  }
  else
    measuring = 0;
} /* ends while measuring */
dist5 = (dist5*14)/10;
if(dist5 <= result)
```

```
    result = dist5;

    /* right and up */
measuring = 1;
i = a;
j = b;
while(measuring){
    j++;
    i--;
    if(j <=cols-1 && i>=0){
      if(the_image[i][j] == value)
        dist6++;
      else
        measuring = 0;
  }
  else
    measuring = 0;
} /* ends while measuring */
dist6 = (dist6*14)/10;
if(dist6 <= result)
    result = dist6;

    /* right and down */
measuring = 1;
i = a;
j = b;
while(measuring){
    j++;
    i++;
    if(j <=cols-1 && i<=rows-1){
      if(the_image[i][j] == value)
        dist7++;
      else
        measuring = 0;
  }
  else
        measuring = 0;
} /* ends while measuring */
dist7 = (dist7*14)/10;
if(dist7 <= result)
    result = dist7;

    /* left and down */
measuring = 1;
i = a;
```

```
   j = b;
   while(measuring){
       j--;
       i++;
       if(j >=0 && i<=rows-1){
           if(the_image[i][j] == value)
               dist8++;
       else
           measuring = 0;
       }
       else
           measuring = 0;
   } /* ends while measuring */
   dist8 = (dist8*14)/10;
   if(dist8 <= result)
       result = dist8;

       return(result);
} /* ends distance_8 */

   /*****************************************
```

mat function finds the medial axis transform for objects in an image. The mat are those points that are minimally distant to more than one boundary point.

```
       ***************************************/

mat(the_image, out_image,
   value, rows, cols)
   short **the_image,
        **out_image,
        value;
   long cols, rows;
{
   int a, b, count, i, j, k,
       length, width;
   for(i=0; i<rows; i++)
     for(j=0; j<cols; j++)
       out_image[i][j] = 0;
   printf("\n");
   for(i=0; i<rows; i++){
     if( (i%10) == 0) printf("%3d", i);
     for(j=0; j<cols; j++){
       if(the_image[i][j] == value)
         out_image[i][j] = mat_d(the_image,
                     i, j, value,
                     rows, cols);
```

```
    } /* ends loop over j */
  } /* ends loop over i */
} /* ends mat */
```

/**
mat_d function helps find the medial axis transform. This function measures the distances from the point to a zero pixel in all eight directions. Look for the two shortest distances in the eight distances. If the two shortest distances are equal, then the point in question is minimally distant to more than one boundary point. Therefore, it is on the medial axis, so return a value. Otherwise, return zero.
**/

```
mat_d(the_image, a, b, value, rows, cols)
    int a, b;
    short **the_image, value;
    long cols, rows;
{
    int i, j, measuring;
    short dist1 = 0,
          dist2 = 0,
          dist3 = 0,
          dist4 = 0,
          dist5 = 0,
          dist6 = 0,
          dist7 = 0,
          dist8 = 0,
          min1 = GRAY_LEVELS,
          min2 = GRAY_LEVELS,
          result = 0;
      /* straight up */
  measuring = 1;
  i = a;
  j = b;
  while(measuring){
      i--;
      if(i >= 0){
        if(the_image[i][j] == value)
           dist1++;
        else
           measuring = 0;
      }
    else
        measuring = 0;
  } /* ends while measuring */
```

```
result = dist1;
min1 = dist1;
   /* straight down */
measuring = 1;
i = a;
j = b;
while(measuring){
    i++;
    if(i <= rows-1){
        if(the_image[i][j] == value)
            dist2++;
    else
        measuring = 0;
  }
  else
    measuring = 0;
} /* ends while measuring */
if(dist2 <= result)
    result = dist2;
if(dist2 < min1){
    min2 = min1;
    min1 = dist2;
}
else
    if(dist2 < min2)
        min2 = dist2;

    /* straight left */
measuring = 1;
i = a;
j = b;
while(measuring){
    j--;
    if(j >= 0){
        if(the_image[i][j] == value)
            dist3++;
    else
        measuring = 0;
}
else
    measuring = 0;
} /* ends while measuring */
if(dist3 <= result)
    result = dist3;
```

```
if(dist3 < min1){
   min2 = min1;
   min1 = dist3;
}
else
   if(dist3 < min2)
      min2 = dist3;
   /* straight right */
measuring = 1;
i = a;
j = b;
while(measuring){
   j++;
   if(j <= cols-1){
      if(the_image[i][j] == value)
         dist4++;
    else
       measuring = 0;
   }
   else
      measuring = 0;
} /* ends while measuring */
if(dist4 <= result)
   result = dist4;
if(dist4 < min1){
   min2 = min1;
   min1 = dist4;
}
else
   if(dist4 < min2)
      min2 = dist4;

   /* left and up */
measuring = 1;
i = a;
j = b;
while(measuring){
   j--;
   i--;
   if(j >= 0 && i>=0){
if(the_image[i][j] == value)
   dist5++;
 else
    measuring = 0;
```

```
      }
    else
        measuring = 0;
  } /* ends while measuring */
  dist5 = ((dist5*14)+7)/10;
  if(dist5 <= result)
      result = dist5;
  if(dist5 < min1){
    min2 = min1;
    min1 = dist5;
  }
  else
    if(dist5 < min2)
      min2 = dist5;
    /* right and up */
  measuring = 1;
  i = a;
  j = b;
  while(measuring){
  j++;
  i--;
  if(j <=cols-1 && i>=0){
    if(the_image[i][j] == value)
        dist6++;
   else
      measuring = 0;
    }
    else
        measuring = 0;
  } /* ends while measuring */
  dist6 = ((dist6*14)+7)/10;
  if(dist6 <= result)
      result = dist6;
  if(dist6 < min1){
    min2 = min1;
    min1 = dist6;
  }
  else
    if(dist6 < min2)
      min2 = dist6;
    /* right and down */
  measuring = 1;
  i = a;
  j = b;
```

```
while(measuring){
    j++;
    i++;
    if(j <=cols-1 && i<=rows-1){
        if(the_image[i][j] == value)
            dist7++;
        else
            measuring = 0;
    }
    else
        measuring = 0;
} /* ends while measuring */
dist7 = ((dist7*14)+7)/10;
if(dist7 <= result)
    result = dist7;
if(dist7 < min1){
    min2 = min1;
    min1 = dist7;
}
else
    if(dist7 < min2)
        min2 = dist7;
    /* left and down */
measuring = 1;
i = a;
j = b;
while(measuring){
    j--;
    i++;
    if(j >=0 && i<=rows-1){
        if(the_image[i][j] == value)
            dist8++;
        else
            measuring = 0;
    }
    else
        measuring = 0;
} /* ends while measuring */
dist8 = ((dist8*14)+7)/10;
if(dist8 <= result)
    result = dist8;
if(dist8 < min1){
    min2 = min1;
    min1 = dist8;
```

```
            }
            else
               if(dist8 < min2)
                   min2 = dist8;
            if(min1 == min2)
                result = value;
            else
                result = 0;
            if(min1 == 0)
                result = 0;
            return(result);
         } /* ends mat_d */
```

```
         /****************************************
```
Thinning function Raster scan the image left to right and examine and thin the left edge pixels (a 0 to value transition). Process them normally and "save" the result. Next, raster scan the image right to left and save. Raster scan top to bottom and save. Raster scan bottom to top and save. That is one complete pass. Keep track of pixels thinned for a pass and quit when you make a complete pass without thinning any pixels.
```
         ****************************************/
```

```
thinning(in_name, out_name, the_image, out_image,
        il, ie, ll, le, value, threshold, once_only)
    char in_name[], out_name[];
    int il, ie, ll, le, once_only;
    short the_image[ROWS][COLS],
          out_image[ROWS][COLS],
          threshold, value;
{
    int    a, b, big_count, count, i, j, k,
           not_finished;

    create_file_if_needed(in_name, out_name, out_image);

    read_tiff_image(in_name, the_image, il, ie, ll, le);
    for(i=0; i<ROWS; i++)
        for(j=0; j<COLS; j++)
            out_image[i][j] = the_image[i][j];
    not_finished = 1;
    while(not_finished){
        if(once_only == 1)
            not_finished = 0;
        big_count = 0;
```

```
/***************************
      Scan left to right Look for 0-value transition
***************************/
printf("\n");
for(i=1; i<ROWS-1; i++){
   if( (i%10) == 0) printf("%3d", i);
   for(j=1; j<COLS-1; j++){
      if(the_image[i][j-1] == 0 &&
         the_image[i][j] == value){
         count = 0;
         for(a=-1; a<=1; a++){
            for(b=-1; b<=1; b++){
               if(the_image[i+a][j+b] == 0)
                  count++;
            } /* ends loop over b */
         } /* ends loop over a */
         if(count > threshold){
            if(can_thin(the_image, i, j, value)){
               out_image[i][j] = 0;
               big_count++;
            } /* ends if can_thin */
         } /* ends if count > threshold */
      } /* ends if the_image == value */
   } /* ends loop over j */
} /* ends loop over i */

   /*************************************
   Copy the output back to the input.
   *************************************/
for(i=0; i<ROWS; i++)
   for(j=0; j<COLS; j++)
      the_image[i][j] = out_image[i][j];
   /***************************
   Scan right to left
   ***************************/
printf("\n");
for(i=1; i<ROWS-1; i++){
   if( (i%10) == 0) printf("%3d", i);
   for(j=1; j<COLS-1; j++){
      if(the_image[i][j+1] == 0 &&
         the_image[i][j] == value){
         count = 0;
         for(a=-1; a<=1; a++){
            for(b=-1; b<=1; b++){
```

```
                    if(the_image[i+a][j+b] == 0)
                          count++;
              } /* ends loop over b */
        } /* ends loop over a */
        if(count > threshold){
           if(can_thin(the_image, i, j, value)){
             out_image[i][j] = 0;
             big_count++;
             } /* ends if can_thin */
           } /* ends if count > threshold */
         } /* ends if the_image == value */
      } /* ends loop over j */
   } /* ends loop over i */
    /*************************************
   *Copy the output back to the input.
   *************************************/
for(i=0; i<ROWS; i++)
   for(j=0; j<COLS; j++)
     the_image[i][j] = out_image[i][j];

    /**************************
   Scan top to bottom
   **************************/
printf("\n");
for(j=1; j<COLS-1; j++){
  if( (j%10) == 0) printf("%3d", j);
  for(i=1; i<ROWS-1; i++){
   if(the_image[i-1][j] == 0 &&
      the_image[i][j] == value){
      count = 0;
     for(a=-1; a<=1; a++){
         for(b=-1; b<=1; b++){
            if(the_image[i+a][j+b] == 0)
                count++;
        } /* ends loop over b */
      } /* ends loop over a */
      if(count > threshold){
       if(can_thin(the_image, i, j, value)){
          out_image[i][j] = 0;
          big_count++;
        } /* ends if can_thin */
        } /* ends if count > threshold */
      } /* ends if the_image == value */
```

```
  } /* ends loop over i */
} /* ends loop over j */

  /************************************
       Copy the output back to the input.
  ************************************/

for(i=0; i<ROWS; i++)
   for(j=0; j<COLS; j++)
      the_image[i][j] = out_image[i][j];

  /*************************
       Scan bottom to top
  *************************/
printf("\n");
for(j=1; j<COLS-1; j++){
   if( (j%10) == 0) printf("%3d", j);
   for(i=1; i<ROWS-1; i++){
     if(the_image[i+1][j] == 0 &&
        the_image[i][j] == value){
        count = 0;
     for(a=-1; a<=1; a++){
        for(b=-1; b<=1; b++){
           if(the_image[i+a][j+b] == 0)
              count++;
        } /* ends loop over b */
      } /* ends loop over a */
      if(count > threshold){
        if(can_thin(the_image, i, j, value)){
           out_image[i][j] = 0;
           big_count++;
        } /* ends if can_thin */
      } /* ends if count > threshold */
     } /* ends if the_image == value */
   } /* ends loop over i */
} /* ends loop over j */

  /************************************
       Copy the output back to the input.
  ************************************/

for(i=0; i<ROWS; i++)
   for(j=0; j<COLS; j++)
      the_image[i][j] = out_image[i][j];
      printf("\n\nThinned %d pixels", big_count);
   if(big_count == 0)
```

```
          not_finished = 0;
      else{
        for(i=0; i<ROWS; i++)
           for(j=0; j<COLS; j++)
              the_image[i][j] = out_image[i][j];
        } /* ends else */
    } /* ends while not_finished */
    fix_edges(out_image, 3);
    write_array_into_tiff_image(out_name, out_image,
                      il, ie, ll, le);
} /* ends thinning */


can_thin(the_image, i, j, value)
    int i, j;
    short the_image[ROWS][COLS], value;
{
    int a, b, c, d, count,
        no_neighbor, one=1, zero=0;
    short temp[3][3];
    for(a=-1; a<2; a++)
      for(b=-1; b<2; b++)
         temp[a+1][b+1] = the_image[i+a][b+j];

temp[1][1] = 0;
for(a=0; a<3; a++){
    for(b=0; b<3; b++){
       if(temp[a][b] == value){
          temp[a][b] = 0;
 no_neighbor = 1;
  for(c=-1; c<2; c++){
    for(d=-1; d<2; d++){
        if( ((a+c) >= 0) &&
          ((a+c) <= 2) &&
          ((b+d) >= 0) &&
          ((b+d) <= 2)){
         if(temp[a+c][b+d] == value){
           no_neighbor = 0;
           } /* ends if temp == value */
          } /* ends if part of temp array */
       } /* ends loop over d */
     } /* ends loop over c */
    temp[a][b] = value;
    if(no_neighbor){
```

```
      return(zero);
      }
   } /* ends if temp[a][b] == value */
   } /* ends loop over b */
 } /* ends loop over a */

 return(one);
} /* ends can_thin */

dilate_not_join(in_name, out_name, the_image, out_image,
          il, ie, ll, le, value, threshold)
   char in_name[], out_name[];
   int il, ie, ll, le;
   short the_image[ROWS][COLS],
         out_image[ROWS][COLS],
         threshold, value;
{
   int a, b, count, i, j, k;

   create_file_if_needed(in_name, out_name, out_image);

   read_tiff_image(in_name, the_image, il, ie, ll, le);

   for(i=0; i<ROWS; i++)
     for(j=0; j<COLS; j++)
       out_image[i][j] = the_image[i][j];

     printf("\n");
     for(i=1; i<ROWS-1; i++){
       if( (i%10) == 0) printf("%3d", i);
       for(j=1; j<COLS-1; j++){
         if(the_image[i][j-1] == value &&
           the_image[i][j] == 0){
           count = 0;
           for(a=-1; a<=1; a++){
             for(b=-1; b<=1; b++){
               if(the_image[i+a][j+b]==value)
                  count++;
             } /* ends loop over b */
           } /* ends loop over a */
           if(count > threshold){
             if(can_dilate(the_image,i,j,value)){
               out_image[i][j] = value;
             } /* ends if can_dilate */
           } /* ends if count > threshold */
         } /* ends if the_image == value */
```

```
    } /* ends loop over j */
} /* ends loop over i */

for(i=0; i<ROWS; i++)
   for(j=0; j<COLS; j++)
     the_image[i][j] = out_image[i][j];
 printf("\n");
 for(i=1; i<ROWS-1; i++){
    if( (i%10) == 0) printf("%3d", i);
    for(j=1; j<COLS-1; j++){
       if(the_image[i][j+1] == value &&
         the_image[i][j] == 0){
         count = 0;
         for(a=-1; a<=1; a++){
            for(b=-1; b<=1; b++){
               if(the_image[i+a][j+b]==value)
                  count++;
             } /* ends loop over b */
       } /* ends loop over a */
       if(count > threshold){
         if(can_dilate(the_image,i,j,value)){
            out_image[i][j] = value;
         } /* ends if can_dilate */
        } /* ends if count > threshold */
    } /* ends if the_image == value */
  } /* ends loop over j */
} /* ends loop over i */
  /************************************
  *
  * Copy the output back to the input.
  *
  ************************************/

for(i=0; i<ROWS; i++)
   for(j=0; j<COLS; j++)
     the_image[i][j] = out_image[i][j];

printf("\n");
for(j=1; j<COLS-1; j++){
   if( (j%10) == 0) printf("%3d", j);
   for(i=1; i<ROWS-1; i++){
     if(the_image[i-1][j] == value &&
       the_image[i][j] == 0){
       count = 0;
       for(a=-1; a<=1; a++){
```

```
        for(b=-1; b<=1; b++){
            if(the_image[i+a][j+b]==value)
                count++;
        } /* ends loop over b */
    } /* ends loop over a */
  if(count > threshold){
    if(can_dilate(the_image,i,j,value)){
      out_image[i][j] = value;
      } /* ends if can_dilate */
     } /* ends if count > threshold */
    } /* ends if the_image == value */
 } /* ends loop over i */
} /* ends loop over j */

for(i=0; i<ROWS; i++)
   for(j=0; j<COLS; j++)
      the_image[i][j] = out_image[i][j];
printf("\n");
for(j=1; j<COLS-1; j++){
   if( (j%10) == 0) printf("%3d", j);
   for(i=1; i<ROWS-1; i++){
    if(the_image[i+1][j] == value &&
      the_image[i][j] == 0){
      count = 0;
      for(a=-1; a<=1; a++){
          for(b=-1; b<=1; b++){
              if(the_image[i+a][j+b]==value)
                  count++;
          } /* ends loop over b */
        } /* ends loop over a */
    if(count > threshold){
      if(can_dilate(the_image,i,j,value)){
        out_image[i][j] = value;
        } /* ends if can_dilate */
       } /* ends if count > threshold */
      } /* ends if the_image == value */
   } /* ends loop over i */
} /* ends loop over j */
for(i=0; i<ROWS; i++)
   for(j=0; j<COLS; j++)
      the_image[i][j] = out_image[i][j];
fix_edges(out_image, 3);

write_array_into_tiff_image(out_name, out_image,
                   il, ie, ll, le);
```

```
} /* ends dilate_not_join */
  can_dilate(the_image, i, j, value)
  int  i,  j;
  short the_image[ROWS][COLS], value;
{
  int a, b, c, d, count, found=0,
     no_neighbor,
     stack_pointer=-1,
     stack_empty=1,
     stack[12][2],
     pop_a, pop_b,
     one=1,
     zero=0;
  short first_value, label = 2, temp[3][3];
  for(a=-1; a<2; a++)
    for(b=-1; b<2; b++)
      temp[a+1][b+1] = the_image[i+a][b+j];
    for(a=0; a<3; a++){
      for(b=0; b<3; b++){
        stack_empty = 1;
        stack_pointer = -1;
        if(temp[a][b] == value){
           little_label_and_check(temp, stack, label,
&stack_empty,
&stack_pointer,
                       a, b, value);
         found = 1;
     } /* ends if temp == value */

    while(stack_empty == 0){
       pop_a = stack[stack_pointer][0]; /* POP */
       pop_b = stack[stack_pointer][1]; /* POP */
       --stack_pointer;
       if(stack_pointer <= 0){
         stack_pointer = 0;
         stack_empty = 1;
       } /* ends if stack_pointer */
 little_label_and_check(temp, stack, label,
&stack_empty,
&stack_pointer,
                   pop_a, pop_b, value);
     } /* ends while stack_empty == 0 */
     if(found){
        found = 0;
```

```
            label++;
         } /* ends if object_found */
      } /* ends loop over b */
} /* ends loop over a */

    first_value = -1;
for(a=0; a<3; a++){
   for(b=0; b<3; b++){
      if(temp[a][b] != 0 &&
         first_value == -1){
         first_value = temp[a][b];
      }
     if(temp[a][b] != 0 &&
        first_value != -1){
        if(temp[a][b] != first_value){
          return(zero);
        }
      }
   } /* ends loop over b */
} /* ends loop over a */

return(one);
} /* ends can_dilate */

 little_label_and_check(temp, stack, label, stack_empty,
               stack_pointer, a, b, value)
 int a, b, stack[12][2],
    *stack_empty, *stack_pointer;
 short temp[3][3], label, value;
{
 int c, d;

temp[a][b] = label;
for(c=a-1; c<=a+1; c++){
   for(d=b-1; d<=b+1; d++){
     if(c >= 0 &&
        c <= 2 &&
        d >= 0 &&
        d <= 2)
        if(temp[c][d] == value){ /* PUSH */
           *stack_pointer = *stack_pointer + 1;
           stack[*stack_pointer][0] = c;
           stack[*stack_pointer][1] = d;
           *stack_empty = 0;
        } /* ends if temp == value */
     } /* ends loop over d */
```

```
   } /* ends loop over c */

} /* ends little_label_and_check */
special_closing(in_name, out_name, the_image,
        out_image, il, ie, ll, le,
        value, threshold, number)
  char in_name[], out_name[];
  int il, ie, ll, le, number;
  short the_image[ROWS][COLS],
      out_image[ROWS][COLS],
      threshold, value;
{
  int a, b, count, i, j, k;
  create_file_if_needed(in_name, out_name, out_image);
  read_tiff_image(in_name, the_image, il, ie, ll, le);
  dilate_not_join(in_name, out_name, the_image,
            out_image, il, ie, ll, le,
            value, threshold);
  if(number > 1){
    count = 1;
    while(count < number){
      count++;
      dilate_not_join(out_name, out_name, the_image,
            out_image, il, ie, ll, le,
            value, threshold);
    } /* ends while */
  } /* ends if number > 1 */
 erosion(out_name, out_name, the_image,
        out_image, il, ie, ll, le,
        value, threshold);
  write_array_into_tiff_image(out_name, out_image,
            il, ie, ll, le);
} /* ends special_closing */

special_opening(in_name, out_name, the_image,
        out_image, il, ie, ll, le,
        value, threshold, number)
  char in_name[], out_name[];
  int il, ie, ll, le, number;
  short the_image[ROWS][COLS],
      out_image[ROWS][COLS],
      threshold, value;
{
  int a, b, count, i, j, k;
```

```
   create_file_if_needed(in_name, out_name, out_image);
   read_tiff_image(in_name, the_image, il, ie, ll, le);
   thinning(in_name, out_name, the_image,
         out_image, il, ie, ll, le,
         value, threshold, 1);
  if(number > 1){
    count = 1;
    while(count < number){
      count++;
      thinning(out_name, out_name, the_image,
           out_image, il, ie, ll, le,
           value, threshold, 1);
    } /* ends while */
  } /* ends if number > 1 */
 dilation(out_name, out_name, the_image,
         out_image, il, ie, ll, le,
         value, threshold);
 write_array_into_tiff_image(out_name, out_image,
                    il, ie, ll, le);
} /* ends special_opening */
edm(in_name, out_name, the_image, out_image,
       il, ie, ll, le, value)
  char in_name[], out_name[];
  int il, ie, ll, le;
  short the_image[ROWS][COLS],
      out_image[ROWS][COLS],
      value;
{
  int a, b, count, i, j, k;

  create_file_if_needed(in_name, out_name, out_image);

  read_tiff_image(in_name, the_image, il, ie, ll, le);

  for(i=0; i<ROWS; i++)
    for(j=0; j<COLS; j++)
      out_image[i][j] = 0;

printf("\n");
for(i=0; i<ROWS; i++){
   if( (i%10) == 0) printf("%3d", i);
   for(j=0; j<COLS; j++){
     if(the_image[i][j] == value)
        out_image[i][j] = distance_8(the_image,
                       i, j, value);
```

```
      } /* ends loop over j */
   } /* ends loop over i */

   write_array_into_tiff_image(out_name, out_image,
                        il, ie, ll, le);

   } /* ends edm */

   distance_8(the_image, a, b, value)
     int a, b;
     short the_image[ROWS][COLS], value;
   {
      int i, j, measuring;
      short dist1 = 0,
            dist2 = 0,
            dist3 = 0,
            dist4 = 0,
            dist5 = 0,
            dist6 = 0,
            dist7 = 0,
            dist8 = 0,
            result = 0;

         /* straight up */
      measuring = 1;
      i = a;
      j = b;
      while(measuring){
         i--;
         if(i >= 0){
            if(the_image[i][j] == value)
              dist1++;
         else
           measuring = 0;
         }
       else
          measuring = 0;
      } /* ends while measuring */
      result = dist1;

         /* straight down */
      measuring = 1;
      i = a;
      j = b;
      while(measuring){
         i++;
```

```
    if(i <= ROWS-1){
       if(the_image[i][j] == value)
          dist2++;
     else
       measuring = 0;
       }
     else
       measuring = 0;
} /* ends while measuring */
if(dist2 <= result)
   result = dist2;

   /* straight left */
measuring = 1;
i = a;
j = b;
 while(measuring){
    j--;
    if(j >= 0){
       if(the_image[i][j] == value)
          dist3++;
     else
        measuring = 0;
     }
    else
      measuring = 0;
} /* ends while measuring */
if(dist3 <= result)
   result = dist3;
   /* straight right */
measuring = 1;
i = a;
j = b;
 while(measuring){
    j++;
    if(j <= COLS-1){
       if(the_image[i][j] == value)
          dist4++;
     else
        measuring = 0;
     }
    else
      measuring = 0;
} /* ends while measuring */
```

```
if(dist4 <= result)
    result = dist4;
    /* left and up */
measuring = 1;
i = a;
j = b;
while(measuring){
  j--;
  i--;
  if(j >= 0 && i>=0){
    if(the_image[i][j] == value)
        dist5++;
    else
        measuring = 0;
  }
  else
      measuring = 0;
} /* ends while measuring */
dist5 = (dist5*14)/10;
if(dist5 <= result)
    result = dist5;
    /* right and up */
measuring = 1;
i = a;
j = b;
while(measuring){
  j++;
  i--;
  if(j <=COLS-1 && i>=0){
    if(the_image[i][j] == value)
      dist6++;
    else
      measuring = 0;
  }
  else
    measuring = 0;
} /* ends while measuring */
dist6 = (dist6*14)/10;
if(dist6 <= result)
    result = dist6;
    /* right and down */
measuring = 1;
i = a;
j = b;
```

```
  while(measuring){
     j++;
     i++;
     if(j <=COLS-1 && i<=ROWS-1){
       if(the_image[i][j] == value)
          dist7++;
     else
       measuring = 0;
  }
 else
    measuring = 0;
} /* ends while measuring */
dist7 = (dist7*14)/10;
if(dist7 <= result)
   result = dist7;

  /* left and down */
measuring = 1;
i = a;
j = b;
  while(measuring){
     j--;
     i++;
     if(j >=0 && i<=ROWS-1){
        if(the_image[i][j] == value)
        dist8++;
     else
        measuring = 0;
  }
   else
      measuring = 0;
} /* ends while measuring */
dist8 = (dist8*14)/10;
if(dist8 <= result)
   result = dist8;

  return(result);
} /* ends distance_8 */

mat(in_name, out_name, the_image, out_image,
    il, ie, ll, le, value)
    char in_name[], out_name[];
    int il, ie, ll, le;
    short the_image[ROWS][COLS],
          out_image[ROWS][COLS],
```

```
            value;
{
  int a, b, count, i, j, k,
      length, width;

  create_file_if_needed(in_name, out_name, out_image);

  read_tiff_image(in_name, the_image, il, ie, ll, le);

  for(i=0; i<ROWS; i++)
    for(j=0; j<COLS; j++)
      out_image[i][j] = 0;
  printf("\n");

  for(i=0; i<ROWS; i++){
    if( (i%10) == 0) printf("%3d", i);
    for(j=0; j<COLS; j++){
      if(the_image[i][j] == value)
        out_image[i][j] = mat_d(the_image,
                            i, j, value);
    } /* ends loop over j */
  } /* ends loop over i */

  write_array_into_tiff_image(out_name, out_image,
                      il, ie, ll, le);
} /* ends mat */
    /*******************************************
```

mat_d function helps find the medial axis transform. This function measures the distances from the point to a zero pixel in all eight directions. Look for the two shortest distances in the eight distances.

```
    *******************************************/

mat_d(the_image, a, b, value)
  int a, b;
  short the_image[ROWS][COLS], value;
{
  int i, j, measuring;
  short dist1 = 0,
        dist2 = 0,
        dist3 = 0,
        dist4 = 0,
        dist5 = 0,
        dist6 = 0,
        dist7 = 0,
        dist8 = 0,
        min1 = 255,
```

```
        min2 = 255,
        result = 0;
    /* straight up */
measuring = 1;
i = a;
j = b;
  while(measuring){
    i--;
    if(i >= 0){
       if(the_image[i][j] == value)
         dist1++;
      else
        measuring = 0;
  }
  else
     measuring = 0;
} /* ends while measuring */
result = dist1;
min1 = dist1;

    /* straight down */
measuring = 1;
i = a;
j = b;
while(measuring){
    i++;
    if(i <= ROWS-1){
       if(the_image[i][j] == value)
         dist2++;
    else
      measuring = 0;
  }
  else
     measuring = 0;
} /* ends while measuring */
if(dist2 <= result)
   result = dist2;
if(dist2 < min1){
   min2 = min1;
   min1 = dist2;
}
else
   if(dist2 < min2)
     min2 = dist2;
```

```
     /* straight left */
measuring = 1;
i = a;
j = b;
 while(measuring){
     j--;
     if(j >= 0){
        if(the_image[i][j] == value)
           dist3++;
        else
           measuring = 0;
     }
     else
        measuring = 0;
} /* ends while measuring */
if(dist3 <= result)
   result = dist3;
if(dist3 < min1){
   min2 = min1;
   min1 = dist3;
}
else
   if(dist3 < min2)
      min2 = dist3;
   /* straight right */
measuring = 1;
i = a;
j = b;
while(measuring){
     j++;
     if(j <= COLS-1){
        if(the_image[i][j] == value)
        dist4++;
   else
      measuring = 0;
   }
   else
      measuring = 0;
} /* ends while measuring */
if(dist4 <= result)
    result = dist4;
if(dist4 < min1){
   min2 = min1;
   min1 = dist4;
```

```
}
else
   if(dist4 < min2)
      min2 = dist4;

  /* left and up */
measuring = 1;
i = a;
j = b;
while(measuring){
   j--;
   i--;
   if(j >= 0 && i>=0){
      if(the_image[i][j] == value)
         dist5++;
    else
      measuring = 0;
   }
   else
      measuring = 0;
} /* ends while measuring */
dist5 = ((dist5*14)+7)/10;
if(dist5 <= result)
   result = dist5;
if(dist5 < min1){
   min2 = min1;
   min1 = dist5;
}
else
   if(dist5 < min2)
      min2 = dist5;

   /* right and up */
measuring = 1;
i = a;
j = b;
while(measuring){
   j++;
   i--;
   if(j <=COLS-1 && i>=0){
      if(the_image[i][j] == value)
         dist6++;
    else
      measuring = 0;
   }
```

```
      else
         measuring = 0;
   } /* ends while measuring */
   dist6 = ((dist6*14)+7)/10;
   if(dist6 <= result)
         result = dist6;
   if(dist6 < min1){
      min2 = min1;
      min1 = dist6;
   }
   else
      if(dist6 < min2)
         min2 = dist6;

   /* right and down */
   measuring = 1;
   i = a;
   j = b;
   while(measuring){
         j++;
         i++;
         if(j <=COLS-1 && i<=ROWS-1){
            if(the_image[i][j] == value)
               dist7++;
         else
            measuring = 0;
      }
      else
         measuring = 0;
   } /* ends while measuring */
   dist7 = ((dist7*14)+7)/10;
   if(dist7 <= result)
         result = dist7;
   if(dist7 < min1){
      min2 = min1;
      min1 = dist7;
   }
   else
      if(dist7 < min2)
         min2 = dist7;

   /* left and down */
   measuring = 1;
   i = a;
   j = b;
```

```
while(measuring){
   j--;
   i++;
   if(j >=0 && i<=ROWS-1){
     if(the_image[i][j] == value)
        dist8++;
   else
      measuring = 0;
}
else
   measuring = 0;
} /* ends while measuring */
dist8 = ((dist8*14)+7)/10;
if(dist8 <= result)
   result = dist8;
if(dist8 < min1){
   min2 = min1;
   min1 = dist8;
}
else
   if(dist8 < min2)
      min2 = dist8;

if(min1 == min2)
   result = value;
else
   result = 0;

if(min1 == 0)
   result = 0;
   return(result);
} /* ends mat_d */
```

B.12 Morphological Operations Erosion, Dilation, Outlining, Opening and Closing Operations

```
short edmask1[3][3] = {{0, 1, 0},
             {0, 1, 0},
             {0, 1, 0}};
short edmask2[3][3] = {{0, 0, 0},
             {1, 1, 1},
             {0, 0, 0}};
short edmask3[3][3] = {{0, 1, 0},
             {1, 1, 1},
```

```
                {0, 1, 0}};
short edmask4[3][3] = {{1, 1, 1},
                {1, 1, 1},
                {1, 1, 1}};
```

```
    /*********************************************
```
erosion function performs the erosion operation. If a value pixel has more than the threshold number of 0 neighbors, then erode it by setting it to 0.
```
    *********************************************/
```

```
erosion(the_image, out_image,
    value, threshold,
    rows, cols)
  int threshold;
  short **the_image,
    **out_image,
    value;
  long cols, rows;
{
  int a, b, count, i, j, k;

/***************************
* Loop over image array
***************************/

for(i=0; i<rows; i++)
   for(j=0; j<cols; j++)
      out_image[i][j] = the_image[i][j];
printf("\n");
for(i=1; i<rows-1; i++){
   if( (i%10) == 0) printf("%3d", i);
   for(j=1; j<cols-1; j++){
      if(the_image[i][j] == value){
         count = 0;
         for(a=-1; a<=1; a++){
            for(b=-1; b<=1; b++){
               if( (i+a) >= 0){
                  if(the_image[i+a][j+b] == 0)
                     count++;
               }
            }
         } /* ends loop over b */
      } /* ends loop over a */
      if(count > threshold){ out_image[i][j] = 0;
      }
   } /* ends if the_image == value */
} /* ends loop over j */
```

```
} /* ends loop over i */
/*****
fix_edges(out_image, 3, rows, cols);
***/
} /* ends erosion */

   /*********************************************
Dilation function performs the dilation operation. If a 0 pixel has more than threshold
number of value neighbors, you dilate it by setting it to value.
*********************************************/

dilation(the_image, out_image,
      value, threshold,
      rows, cols)
   int threshold;
   short **the_image,
      **out_image,
      value;
   long cols, rows;
{
   int a, b, count, i, j, k;
   int three = 3;

      /*************************
      *   Loop over image array
      *************************/

   for(i=0; i<rows; i++)
     for(j=0; j<cols; j++)
       out_image[i][j] = the_image[i][j];
   printf("\n");
   for(i=1; i<rows-1; i++){
     if( (i%10) == 0) printf("%3d", i);
     for(j=1; j<cols-1; j++){
       out_image[i][j] = the_image[i][j];
       if(the_image[i][j] == 0){
         count = 0;
         for(a=-1; a<=1; a++){
             for(b=-1; b<=1; b++){
                 if(a!=0 && b!=0){
                     if(the_image[i+a][j+b] == value)
                         count++;
                 } /* ends avoid the center pixel */
             } /* ends loop over b */
         } /* ends loop over a */
```

```
            if(count > threshold)
                out_image[i][j] = value;
            } /* ends if the_image == 0 */
        } /* ends loop over j */
    } /* ends loop over i */
    /*****
    fix_edges(out_image, three, rows, cols);
    ***/
} /* ends dilation */
```

    /***
mask_dilation function performs the dilation operation using the erosion-dilation
3x3 masks given above. It works on 0-value images.
    ***/

```
mask_dilation(the_image, out_image,
        value, mask_type,
        rows, cols)
    int mask_type;
    short **the_image,
        **out_image,
        value;
    long cols, rows;
{
        int a, b, count, i, j, k;
        short mask[3][3], max;
```

    /***********************************
Copy the 3x3 erosion-dilation mask specified by the mask_type.
    ***********************************/

```
        switch(mask_type){
            case 1:
                copy_3_x_3(mask, edmask1);
                break;
            case 2:
                copy_3_x_3(mask, edmask2);
                break;
            case 3:
                copy_3_x_3(mask, edmask3);
                break;
            case 4:
                copy_3_x_3(mask, edmask4);
                break;
            default:
```

```
                    printf("\nInvalid mask type, using mask 4");
                    copy_3_x_3(mask, edmask4);
                    break;
}

    /**************************
    *   Loop over image array
    ***************************/

printf("\n");
    for(i=1; i<rows-1; i++){
      if( (i%10) == 0) printf("%3d", i);
      for(j=1; j<cols-1; j++){
        max = 0;
        for(a=-1; a<=1; a++){
          for(b=-1; b<=1; b++){
              if(mask[a+1][b+1] == 1){
                 if(the_image[i+a][j+b] > max)
                     max = the_image[i+a][j+b];
              } /* ends if mask == 1 */
          } /* ends loop over b */
        } /* ends loop over a */
        out_image[i][j] = max;
      } /* ends loop over j */
    } /* ends loop over i */
    /*****
    fix_edges(out_image, 3, rows, cols);
    ***/
} /* ends mask_dilation */

    /*******************************************
mask_erosion function performs the erosion operation using the erosion-dilation
3x3 masks given above. It works on 0-value images.
    *******************************************/

mask_erosion(the_image, out_image,
        value, mask_type,
        rows, cols)
    int mask_type;
    short **the_image,
        **out_image,
        value;
    long cols, rows;
{
    int a, b, count, i, j, k;
    short mask[3][3], min;
```

```
/***********************************
Copy the 3x3 erosion-dilation mask specified by the mask_type.
************************************/

switch(mask_type){
    case 1:
        copy_3_x_3(mask, edmask1);
        break;
    case 2:
        copy_3_x_3(mask, edmask2);
        break;
    case 3:
        copy_3_x_3(mask, edmask3);
        break;
    case 4:
        copy_3_x_3(mask, edmask4);
        break;
    default:
        printf("\nInvalid mask type, using mask 4");
        copy_3_x_3(mask, edmask4);
        break;
  }

    /**************************
    * Loop over image array
    ***************************/

printf("\n");
    for(i=1; i<rows-1; i++){
        if( (i%10) == 0) printf("%3d", i);
        for(j=1; j<cols-1; j++){
            min = value;
            for(a=-1; a<=1; a++){
                for(b=-1; b<=1; b++){
                    if(mask[a+1][b+1] == 1){
                        if(the_image[i+a][j+b] < min)
                            min = the_image[i+a][j+b];
                        } /* ends if mask == 1 */
                    } /* ends loop over b */
                } /* ends loop over a */
                out_image[i][j] = min;
            } /* ends loop over j */
        } /* ends loop over i */
    /*****
    fix_edges(out_image, 3, rows, cols);
    ***/
```

```
} /* ends mask_erosion */

    /**********************************************
copy_3_x_3(a, b)function copies a 3x3 array of shorts from one array to another.
It copies array b into array a.
    **********************************************/

copy_3_x_3(a, b)
    short a[3][3], b[3][3];
{
    int i, j;
    for(i=0; i<3; i++)
        for(j=0; j<3; j++)
            a[i][j] = b[i][j];
} /* ends copy_3_x_3 */

    /*********************************************
Opening is erosion followed by dilation. This routine will use the mask erosion
and dilation.
    *********************************************/

opening(the_image, out_image,
        value, mask_type, number,
        rows, cols)
    int number;
    int mask_type;
    short **the_image,
        **out_image,
        value;
    long cols, rows;
{
    int a, b, count, i, j, k;
    short mask[3][3], max;

        /************************************
        Copy the 3x3 erosion-dilation mask specified by the mask_type.
        ************************************/

    switch(mask_type){
        case 1:
            copy_3_x_3(mask, edmask1);
            break;
        case 2:
            copy_3_x_3(mask, edmask2);
            break;
        case 3:
            copy_3_x_3(mask, edmask3);
```

```
          break;
        case 4:
          copy_3_x_3(mask, edmask4);
          break;
        default:
          printf("\nInvalid mask type, using mask 4");
          copy_3_x_3(mask, edmask4);
          break;
      }
    for(i=0; i<rows; i++)
       for(j=0; j<cols; j++)
          out_image[i][j] = the_image[i][j];

    mask_erosion(the_image, out_image,
               value, mask_type,
               rows, cols);
    if(number > 1){
       count = 1;
       while(count < number){
          count++;
          mask_erosion(the_image, out_image,
               value, mask_type,
               rows, cols);
       } /* ends while */
    } /* ends if number > 1 */

    mask_dilation(the_image,
               out_image,
               value, mask_type,
               rows, cols);
} /* ends opening */

/*******************************************
Closing is dilation followed by erosion. This routine will use the mask erosion
and dilation.
*******************************************/

closing(the_image, out_image,
               value, mask_type, number,
               rows, cols)
       int number;
       int mask_type;
       short **the_image,
               **out_image,
               value;
```

```
        long cols, rows;
{
        int a, b, count, i, j, k;
        short mask[3][3], max;
printf("\nCLOSING> value=%d mask=%d number=%d",value,mask_
type,number);

    /***********************************
  Copy the 3x3 erosion-dilation mask specified by the mask_type.
    ***********************************/

 switch(mask_type){
        case 1:
          copy_3_x_3(mask, edmask1);
          break;
        case 2:
          copy_3_x_3(mask, edmask2);
          break;
        case 3:
          copy_3_x_3(mask, edmask3);
          break;
        case 4:
          copy_3_x_3(mask, edmask4);
          break;
        default:
          printf("\nInvalid mask type, using mask 4");
          copy_3_x_3(mask, edmask4);
          break;
 }
 for(i=0; i<rows; i++)
     for(j=0; j<cols; j++)
         out_image[i][j] = the_image[i][j];

 mask_dilation(the_image, out_image,
             value, mask_type,
             rows, cols);

 if(number > 1){
     count = 1;
     while(count < number){
        count++;
        mask_dilation(the_image, out_image,
            value, mask_type,
            rows, cols);
     } /* ends while */
```

```
    } /* ends if number > 1 */
    mask_erosion(the_image, out_image,
                value, mask_type,
                rows, cols);
    } /* ends closing */

    /******************************************
interior_outline function produces the outline of any "holes" inside an object. The
method is: output = erosion of input
    final output = input - output
    ******************************************/

interior_outline(the_image, out_image,
                value, mask_type,
                rows, cols)
    int mask_type;
    short **the_image,
        **out_image,
        value;
    long cols, rows;
{
    int a, b, count, i, j, k;
    short mask[3][3], max;

        /***************************************
    Copy the 3x3 erosion-dilation mask specified by the mask_type.
        ***************************************/

switch(mask_type){
    case 1:
        copy_3_x_3(mask, edmask1);
        break;
    case 2:
        copy_3_x_3(mask, edmask2);
        break;
    case 3:
        copy_3_x_3(mask, edmask3);
        break;
    case 4:
        copy_3_x_3(mask, edmask4);
        break;
        default:
        printf("\nInvalid mask type, using mask 4");
        copy_3_x_3(mask, edmask4);
        break;
}
```

```
   mask_erosion(the_image,
                out_image,
                value, mask_type,
                rows, cols);
      for(i=0; i<rows; i++)
         for(j=0; j<cols; j++)
            the_image[i][j] =
               the_image[i][j] - out_image[i][j];
            for(i=0; i<rows; i++)
            for(j=0; j<cols; j++)
      out_image[i][j] = the_image[i][j];
} /* ends interior_outline */

   /*******************************************
exterior_outline function produces the outline of exterior of an object. The method
is output = dilation of input
   final output = output - input
   *******************************************/

exterior_outline(the_image, out_image,
                value, mask_type,
                rows, cols)
      int mask_type;
      short **the_image,
         **out_image,
         value;
      long cols, rows;
{
      int a, b, count, i, j, k;
      short mask[3][3], max;

   /*************************************
Copy the 3x3 erosion-dilation mask specified by the mask_type.
   *************************************/

switch(mask_type){
    case 1:
       copy_3_x_3(mask, edmask1);
       break;
    case 2:
       copy_3_x_3(mask, edmask2);
       break;
    case 3:
       copy_3_x_3(mask, edmask3);
       break;
```

```
      case 4:
        copy_3_x_3(mask, edmask4);
        break;
      default:
        printf("\nInvalid mask type, using mask 4");
        copy_3_x_3(mask, edmask4);
        break;
   }
   mask_dilation(the_image, out_image,
              value, mask_type,
              rows, cols);

   for(i=0; i<rows; i++)
     for(j=0; j<cols; j++)
       the_image[i][j] =
           out_image[i][j] - the_image[i][j];

   for(i=0; i<rows; i++)
     for(j=0; j<cols; j++)
       out_image[i][j] = the_image[i][j];
   } /* ends exterior_outline */
```

Appendix C

Glossary of Image Processing Terms

A

Adaptive Filter

A filter whose behaviour changes according to image properties.

Addition

A point process that blends the values of corresponding pixels in two input images. A single parameter controls which input image dominates the output image.

Additive Primary Color

The colors red, green and blue which can produce all colours when mixed.

Affine Transformation

A first order geometric transformation that involves a combination of translation, rotation, scaling and skewing. An affine transformation preserves points, straight lines, and planes. Sets of parallel lines remain parallel after an affine transformation.

Area Processes

A category of image-processing techniques that calculate the value of each output-image pixel from the corresponding input-image pixel and its neighbors. Examples include halftoning, sharpening and median filtering.

Artifacts

Unwanted blemishes, which may have been introduced to an image by electrical noise during scanning.

Aspect Ratio

The ratio of the width of an image to its height. Examples include 4:3, the aspect ratio of a standard TV, and 16:9, the aspect ratio of a widescreen TV.

B

Bandwidth

A measure of data speed in bits per second.

Backward Mapping

The technique geometric processes use to calculate output-image pixel values from input-image pixel values. Backward-mapping processes start with a pixel in the output image, find the corresponding pixel in the input image, calculate a new pixel value, and assign the new value to the output-image pixel.

Bayer Filter

Enables a CCD to capture color by covering 2x2 blocks of CCD-sensor elements with one red, one blue and two green filters, which makes color CCDs more sensitive to the green wavelengths of light, just like our eyes.

Bi-linear Interpolation

A method of calculating the value of a pixel with fractional co-ordinates that lies in between a 2x2 neighborhood of pixels.

Binary image

An image in which pixel values may be either 0 or 1.

Bit depth

Number of bits used to describe the color of each pixel.

Blurring

An area process that produces an effect similar to an out-of-focus photograph. Blurring removes the detail in an image by making each pixel more like its neighbors.

Bounding Box

The smallest rectangle that encloses a shape so that each of the four sides touches an extremity of the shape.

Brightness

Determines the intensity of the color presented by a pixel in a color image, or the shade of gray presented by a pixel in a grayscale image.

Brightness Adaptation

The human visual system has the ability to operate over a wide range of illumination levels. The process that allows change in the sensitivity of visual system is known as brightness adaptation.

Brightness Transformation

A point process that maps input brightnesses onto output brightnesses with a linear or non-linear mathematical function.

C

Capture

The process of measuring and visualizing physical phenomena such as visible and non-visible electromagnetic radiation. Examples include taking a photograph and scanning a document.

CCD

A charge-couple device is an imaging sensor that contains a rectangular grid of light-sensitive capacitors. Each capacitor accumulates a charge proportional to the brightness of the light it detects; the brighter the light, the higher the charge.

Chain Code

A method of coding the boundary information of an object.

Closing

A morphological operation produced by following a dilation by an erosion. Often used for filling holes in bitmap images.

Color Lookup Table

A table containing RGB values for 256 colors. Storage space is saved as an 8-bit number which links each image pixel to RGB value held in the table instead of each pixel holding a 24-bit description of its color.

Color Model

Determines how the color in a digital image is represented numerically. Examples include the RGB and HSB color models.

Composition

A point process that overlays the pixels of a foreground input image onto the pixels of a background input image.

Compression

A process that is applied to a digital image to reduce its file size.

Cone

Type of a cell in human eye that is sensitive to color.

Contrast

The difference between the lightest and darkest regions of an image.

Contrast Expansion

An image-processing technique that re-distributes the brightness in an image to eliminate regions that are either too dark or too light. Examples include basic and ends-in contrast expansion.

Convolution

A method of calculating the new value of a central pixel in a neighborhood by multiplying each pixel in the neighborhood by a corresponding weight; the new value of the central pixel is the sum of the multiplications.

Corrupted Pixel

A pixel value altered by noise.

Cropping

A geometric process that reduces the size of an image by discarding the pixels outside a specified region called the crop selection.

D

Diagonal Axis

The line that runs from the top-left corner of an image to the bottom-right corner, or from the top-right corner to the bottom-left corner.

Digital Camera

An imaging device that focuses visible light onto a CCD.

Digital Image

An image captured by an imaging device and represented in a computer as a rectangular grid of pixels.

Dilation

A morphological operation that increases the size of objects in an image by adding a layer of foreground pixels around each object.

DPI

Dots per inch.

E

Edge

Edges mark the boundaries between the objects in a scene. A large change in pixel brightness over a small number of pixels often indicates the presence of an edge.

Edge Detector

An image-processing routine that flags the large changes in pixel brightness that indicate potential edges. Edge detectors often visualize their results in edge maps. Examples include the Sobel, Prewitt, Kirsch and Laplacian edge detectors.

Edge Direction

The angle that specifies the direction of an edge. The angle is perpendicular to the direction of the large change in brightness that indicates the edge.

Edge Magnitude

A number that represents how confident an edge detector is that it has found an edge in an image.

Edge Map

A grayscale output image that visualizes the magnitude of the edge found at each pixel in an input image; the greater the magnitude, the brighter the corresponding edge-map pixel. Thresholding an edge map highlights the strongest edges.

Edge Mask

A set of convolution weights that highlight the size and direction of the edges in an image.

Electromagnetic Spectrum

The complete range of electromagnetic radiation from short wavelength gamma radiation to long wavelength radio waves.

Erosion

A morphological operation that decreases the size of objects in an image by removing a layer of foreground pixels around each object. Often used for removing projections and blobs in bitmap images.

F

Filter

A mathematical procedure that alters the digital numbers in an image.

Flipping

A geometric process that swaps the pixels in an image across the horizontal, vertical and diagonal axes.

Forward Mapping

The technique point and area processes use to calculate output-image pixel values from input-image pixel values. Forward-mapping processes start with a pixel in the input image, calculate a new pixel value, and assign the new value to the corresponding pixel in the output image.

Fourier Transform

The representation of a signal or image as a sum of complex-valued sinusoids that extend infinitely in time.

Frame Averaging

A point process that removes noise from a series of input images taken of the same subject. Each output-image pixel value is the average of the corresponding input-image pixel values.

G

Gamma Correction

Used to calibrate devices, smoothing out any irregularities between input and output signals.

Gaussian Noise

A form of image noise that adds small positive and negative deviations to the pixels in an image, often caused by the random variations between the elements of a CCD sensor. Plotting the number of occurrences of each deviation on a histogram produces the bell-shaped curve of the normal distribution, which is also called the Gaussian distribution.

Geometric Process

A category of image-processing techniques that change the size and shape of an image rather than its contents. Examples include cropping, scaling and rotation.

Grayscale Image

An image composed of pixels that present shades of gray.

H

Halftoning

An area process that simulates shades of gray in bitmap images with patterns of bitmap pixels. The density of each 2x2-pixel pattern depends on the ratio of black to white bitmap pixels.

Highlights

The range of pixel brightnesses that represent the lighter regions of an image.

High-key Image

An image that represents a naturally light subject.

High-contrast Image

An image with large numbers of pixels in the shadows and highlights.

High Frequency

The high frequency information in an image is represented by large changes in pixel brightness over a small number of pixels.

High-pass Filter

A filter that preserves or amplifies the high frequency information in an image. Sharpening is implemented by a high pass filter.

Histogram

The histogram of an image visualizes the distribution of the brightness in the image by plotting the number of occurrences of each brightness.

Histogram Equalization

An image-processing technique that reveals detail hidden in images with a poorly-distributed range of brightnesses.

Horizontal Axis

The line that runs through the center of an image from the left of the image to the right.

HSB

A color model that represents each color with three numbers that specify the hue (H), the saturation (S) and the brightness (B) of the color.

Hue

The color in the HSB color model.

I

Image

An image records a visual snapshot of the world around us.

Image Processing

The field of computer science that develops techniques for enhancing digital images to make them more enjoyable to look at, and easier to analyze by computers as well as humans.

Imaging Device

A piece of equipment that captures an image. Examples include digital cameras, side-scan sonar systems and scanning electron microscopes.

Impulse Noise

Also called salt and pepper noise, impulse noise introduces very light (salt) and very dark (pepper) pixels that stand out from their neighbors.

Interpolation

A method of creating new pixel values from existing pixel values. Examples include nearest-neighbor and bi-linear interpolation.

Input Image

The image transformed by an image-processing routine.

Inversion

A point process that produces an effect similar to photographic negatives: dark pixels become light and light pixels become dark.

K

Kernel

A rectangular grid of convolution weights.

L

Line Edge

A line chain of pixels that separates a region of light pixels from a region of dark pixels.

Linear Brightness Transformation

A category of brightness transformations that lighten and darken images using mathematical functions with curved graphs.

Look-up Table (LUT)

A data structure that minimizes the number of calculations required to process an image with a point process. The brightness of each output-image pixel is found in a LUT at the entry indexed by the brightness of the corresponding input-image pixel.

Lossless Compression

Compression technique that reduces the file size without any loss of information.

Lossy Compression

Compression technique that reduces the file size by removing information from an image.

Low-contrast Image

An image that uses only a small range of the available brightness. Low-contrast images are mostly dark, mostly dull or mostly light.

Low-key Image

An image that represents a naturally dark subject.

Low Frequency

The low frequency information in an image is represented by small changes in pixel brightness over a small number of pixels.

Low-pass Filter

A filter that discards or attenuates the high frequency information in an image and preserves the low frequency information. Removing the high frequency information from an image removes the detail and blurs the image. Blurring is implemented by a low pass filter.

M

Median Filtering

An area process that removes noise by replacing the central pixel in a neighborhood with the median pixel value of the neighborhood.

Mid-tones

The range of pixel brightnesses that represent the regions of an image in between the shadows and highlights.

Morphological Operation

A category of image-processing techniques that operate on the structure of the objects in an image.

N

Noise

Unwanted changes to the values of the pixels in an image, often introduced by the imaging device during capture. Examples include impulse noise and Gaussian noise.

Nearest-neighbor Interpolation

A method of creating values for pixels with fractional co-ordinates that duplicates the value of the pixel with integer co-ordinates nearest to the fractional co-ordinates.

Neighborhood Averaging

An area process that removes noise by replacing the central pixel in a neighborhood with the average pixel value of the neighborhood.

Non-linear Brightness Transformation

A category of brightness transformations that change the brightness of an image using mathematical functions with straight-line graphs. Examples include inversion and posterization.

Non-primary Color

A color created by mixing the red, green and blue primary colors of the RGB color model.

NTSC Grayscale

A shade of gray produced by multiplying the brightnesses of the RGB components of a color pixel by a set of weights that emphasize the green component. Named after the committee that oversees US television.

O

Opening

A morphological operation produced by following an erosion by a dilation. Often used for filling holes in bitmap images.

Outlying Pixel

A pixel with an extreme brightness that is much higher or lower than the brightnesses of the other pixels in the image.

Output Image

An image that contains the results of applying an image-processing routine to an image.

P

Photo Restoration

The application of a series of image-processing routines to enhance a damaged photograph.

Pixel

A square unit of visual information that represents a tiny part of a digital image.

Pixel Depth

The number of colors or shades of gray a pixel can present. Bitmap pixels have depth two, typical grayscale pixels have depth 256, and typical color pixels have depth 16,777,216.

Pixel Neighborhood

A region of pixels processed by an area process. Typical neighborhood dimensions are 3x3 pixels and 5x5 pixels.

Point Processes

A category of image-processing techniques that calculate the value of each output-image pixel from the value of the corresponding input-image pixel. Examples include inversion and pseudo-color.

Posterization

A linear brightness transformation that reduces the number of brightnesses in an image.

Pseudo-color

A point process that divides the range of brightness in a grayscale input image into groups and assigns each group a color. Each output-image pixel is assigned the color that represents the group into which falls the brightness of the corresponding input-image pixel.

Potential Edge

Edge detectors flag all large changes in pixel brightness over a small number of pixels as a potential edge. An edge-analysis system then decides whether the change in brightness represents the border of an object—a real edge—or some other feature of the object, such as its texture.

Primary Colors

The colors red, green and blue from which all other colors in the RGB color model are mixed.

Psychovisual Redundancy

Psychovisual redundancy is due to the fact that the human eye does not respond with equal intensity to all visual information.

Q

Quantization

The calculation that maps the fractional measurements made by imaging devices onto proportional integer pixel brightnesses.

R

Ramp Edge

A region of pixels that separates a region of light pixels from a region of dark pixels. The pixels in the region change gradually from light to dark.

Raw Color

The color of the pixels in an image captured by a color CCD before the two unknown RGB-component brightnesses of each pixel have been interpolated from the known brightnesses of the corresponding components of neighboring pixels.

Resolution

The number of pixels available to represent the details of the subject of a digital image.

RGB

A color model that represents each color with three numbers that specify the amounts of red (R), green (G) and blue (B) that produce the color.

RGB Color Cube

Visualizes the amounts of red, green and blue required to produce each color in the RGB color model as a point in a cube at co-ordinates (x, y, z).

Rod

The type of cells in human eye which detect variations in brightness and are not sensitive to colors.

Roof Edge

A region of pixels that separates a region of light pixels from a region of dark pixels. The pixels in a roof edge increase in brightness to their maximum at the apex of the roof and then decrease to meet the region of pixels on the other side of the edge.

Rotation

A geometric process that turns an image about its center by a specified angle.

Rotation Hole

An output-image pixel not assigned a value when the input image is rotated with forward mapping.

Row-Column Co-ordinates

The pair of numbers that locate a pixel in the rows and columns of the rectangular grid of pixels that represent a digital image.

S

Sampling

The process of mapping a continuous quantity of electromagnetic radiation, such as light or X-rays falling on a sensor, onto a discrete, rectangular grid of pixels.

Saturation

The component of the HSB color model that controls the amount of white mixed into the hue.

Segmentation

A process to partition an image into group of pixels which are similar based on some criteria.

Scale Factor

A fractional number that controls whether a scaling process enlarges or reduces an image. Scale factors between zero and one reduce images; scale factors greater than one enlarge images.

Scaling

A geometric process that changes the size of an image.

Scanner

An imaging device that focuses light reflected from a document onto a CCD that moves across and down the document.

Scanning Electron Microscope

An imaging device that uses electrons to capture images of microscopic objects. A SEM fires a beam of electrons at the surface of the sample and counts the number of electrons dislodged from the surface by the beam; the greater the number of dislodged electrons, the brighter the corresponding point in the image.

Sequential-index Co-ordinate

The number that locates a pixel in an image when the pixels in the image are laid end to end in a sequence.

Shadows

The range of pixel brightnesses that represent the darker regions of an image.

Sharpening

An area process that emphasizes the detail in an image.

Side-scan Sonar System

An imaging device that uses sound waves to capture images of underwater objects. Side-scan sonar systems measure the strength of the acoustic reflections of sound waves directed at the sea bed; the stronger the acoustic reflection, the brighter the corresponding pixel in the image.

Spatial Domain

Spatial domain refers to the 2D image plane in terms of pixel intensities.

Spatial Resolution

Describes the finest detail visible to human eye.

Step Edge

The ideal edge shape characterized by a large, immediate change in pixel brightness from a region of light pixels to a region of dark pixels.

Structuring Element

The rectangular grid of binary values used by morphological operations to assign a new value to a pixel in a bitmap image.

Subtraction

A point process that identifies the pixels that differ between two input images.

T

Thresholding

A point process that produces a bitmap version of a grayscale image. Black bitmap pixels represent grayscale pixels darker than a threshold brightness; white bitmap pixels represent grayscale pixels lighter than the threshold.

Thermal Camera

An imaging device that measures the amount of infra-red light emitted by the subject of the image. Hotter objects emit more infra-red light, which show up as the brighter regions of a thermal image.

Thermal Image

A grayscale image captured by a thermal camera. Often enhanced with pseudo-color to assign the same color to regions of pixels that represent similar temperatures.

V

Vector Image

An image composed of elements like line, polygon, etc.

Vertical Axis

The line that runs through the center of an image from the top of the image to the bottom.

Wavelet

Oscillatory functions of finite duration.

Wavelet Transform

Basically, the representation of an image using wavelet functions at different locations and scales.

Zooming

A process used to change the size of an image.

Bibliography

Abbasi, S., Mokhtarian, F. and Kittler, J. 2000. Enhancing CSS-based shape retrieval for objects with shallow concavities. Image and Vision Computing 18(3): 199–211.

Aggarwal, C. C., Hinneburg, A. and Keim, D.A. 2001. On the surprising behavior of distance metrics in high dimensional space. Lecture Notes in Computer Science, pp. 420–434.

Ahmed, H. A., El Gayar, N. and Onsi, H. 2008. A New Approach in Content-Based Image Retrieval Using Fuzzy Logic" Proceedings of INFOS'2008.

Alajlan, N., Kamel, M. S. and Freeman, G. 2006. Multi-object image retrieval based on shape and topology. Signal Processing: Image Communication 21: 904–918.

Alajlan, N., Rube, I. E., Kamel, M. S. and Freeman, G. 2007. Shape retrieval using triangle-area representation and dynamic space warping. Pattern Recognition 40(7): 1911–1920.

Arbter, K., Snyder, W. E., Burkhardt, H. and Hirzinger. G. 1990. Application of affine-invariant Fourier descriptors to recognition of 3D objects. IEEE Trans. Pattern Analysis and Machine Intelligence 12: 640–647.

Arbter, K., Snyder, W., Burkhardt, H. and Hirzinger, G. 1990. Applications of affine-invariant Fourier descriptors to recognition of 3-D objects. IEEE Trans. Pattern Analysis and Machine Intelligence 12(7): 640–646.

Arica, N. and Vural, F. 2003. BAS: a perceptual shape descriptor based on the beam angle statistics. Pattern Recognition Letters 24(9-10).

Arkin, E. M., Chew, L. P., Huttenlocher, D. P., Kedem, K. and Mitchell, J. S. B. 1991. An efficiently computable metric for comparing polygonal shapes. IEEE Trans. Pattern Analysis and Machine Intelligence 13(3): 209–226.

Aslandogan, Y. A. and Yu, C. T. 1999. Techniques and Systems for Image and Video Retrieval. IEEE Transactions on Knowledge and Data Engineering 11(1): 56–63.

Assfalg, J., Bimbo, A. D. and Pala. P. 2000. Using multiple examples for content-based retrieval," Proc. Int'l Conf. Multimedia and Expo.

Badawy, O. E. and Kamel, M. 2004. Shape Retrieval using Concavity Trees. pp. 111–114. In: Proceedings of the 17th International Conference on Pattern Recognition.

Ballard, D. H. and Brown, C. M. 1982. Computer Vision, New, Jersey: Prentice Hall.

Banon, G. J. F. et al. 2007. Mathematical morphology and its applications to signal and image processing: 8th International Symposium on Mathematical Morphology. Rio de Janeiro, RJ, Brazil, October 10–13.

Bauckhage, C. and Tsotsos, J. K. 2005. Bounding box splitting for robust shape classification. pp. 478–481. In: Proc. IEEE International Conference on Image Processing.

Beckmann, N. et al. 1990. The R*-tree: An efficient robust access method for points and rectangles. ACM SIGMOD Int. Conf. on Management of Data, Atlantic City, May 1990.

Belongie, S., Malik, J. and Puzicha, J. 2002. Shape Matching and Object Recognition Using Shape Context. IEEE Trans. Pattern Analysis and Machine Intelligence 24(4): 509– 522.

Benson, K. B. 1992. Television Engineering Handbook. McGraw-Hill, London, U.K. 1992.

Berretti, S., Bimbo, A. D. and Pala, P. 2000. Retrieval by shape similarity with perceptual distance and effective indexing. IEEE Trans. on Multimedia 2(4): 225–239.

Bhattacharyya, A. 1943. On a measure of divergence between two statistical populations defined by their probability distributions. Bulletin of the Calcutta Mathematical Society 35: 99–109.

Birren, F. 1969. Munsell: A Grammar of Color. Van Nostrand Reinhold, New York, N.Y..

Blaser, A.1979. Database Techniques for Pictorial Applications, Lecture Notes in Computer Science, Vol.81, Springer Verlag GmbH.

Borgefors, G. 1986. Distance transformations in digital images. Computer Vision, Graphics, And Image Processing 34(3): 344–371.

Borgefors, G.1986. Distance Transformations in Digital Images. Computer Vision, Graphics, and Image Processing, 344–371.

Bovik, A. C. (ed.). 2005. Handbook of Image and Video Processing, second edition, Academic Press, NY.

Boxer, L. and Miller, R. 2000. Efficient computation of the Euclidean distance transform. Computer Vision and Image Understanding 80: 379–383.

Boynton, R. M. 1990. Human Color Vision. Halt, Rinehart and Winston.

Burger, W. and Burge, M. J. 2005. Principles of Digital Image Processing, Springer, London, UK.

Burkhardt, H. and Siggelkow, S. 2000. Invariant features for discriminating between equivalence classes. Nonlinear Model-based Image Video Processing and Analysis, John Wiley and Sons.

Carson, C., Thomas, M., Belongie, S., Hellerstein, J. M. and Malik, J. 1999. Blobworld: A system for region-based image indexing and retrieval," Proceedings of the Third International Conference VISUAL'99, Amsterdam, The Netherlands, Lecture Notes in Computer Science 1614. Springer.

Catalan, J. A. and Jin, J. S. 2000. Dimension reduction of texture features for image retrieval using hybrid associative neural networks. IEEE International Conference on Multimedia and Expo 2: 1211–1214.

Celebi, M. E. and Aslandogan, Y. A. 2005. A Comparative Study of Three Moment-Based Shape Descriptors. pp. 788–793. In: Proc. of the International Conference of Information Technology: Codingand Computing.

Celenk, M. 1990. A color clustering technique for image segmentation. Computer Vision, Graphics, and Image Processing 52: 145–170.

Chakrabarti, K., Binderberger, M., Porkaew, K. and Mehrotra, S. 2009. Similar shape retrieval in MARS. In: Proc. IEEE International Conference on Multimedia and Expo.

Chang, N. S. and Fu, K. S. 1980. Query by pictorial example. IEEE Trans. on Software Engineering, 6(6): 519–524, Nov.

Chang, S. K. and Kunii, T. L. 1981. Pictorial database systems," IEEE Computer Magazine, Vol. 14(11):13–21, Nov. 1981.

Chang, S. K. and Liu, S. H. 1984. Picture indexing and abstraction techniques for pictorial databases, IEEE Trans. Pattern Anal. Mach. Intell. 6(4): 475–483.

Chang, S. K., Shi, Q. Y. and Yan, C. Y. 1987. Iconic indexing by 2-D strings. IEEE Trans. on Pattern Anal. Machine Intell. 9(3): 413–428, May 1987.

Chang, S. K., Yan, C. W., Dimitroff, D. C. and Arndt, T. 1988. An intelligent image database system," IEEE Trans. on Software Engineering 14(5): 681–688, May 1988.

Chang, S. K. and Hsu, A. 1992. Image information systems: where do we go from here? IEEE Trans. On Knowledge and Data Engineering 5(5): 431–442, Oct.1992.

Chang, T. and Kuo, C.C.l. 1993. Texture analysis and classification with tree-structured wavelet transform. IEEE Trans. on Image Processing 2(4): 429–441, October 1993.

Chen, G. and Bui, T. D. 1999. Invariant Fourier-wavelet descriptor for pattern recognition. Pattern Recognition 32: 1083–1088.

Chen, Y., Wang, J. Z. and Krovetz, R. 2003. An unsupervised learning approach to content-based image retrieval, IEEE Proceedings of the International Symposium on Signal Processing and its Applications, pp. 197–200.

Chuang, C.-H. and Kuo, C.-C. 1996. Wavelet Descriptor of Planar Curves: Theory and Applications. IEEE Trans. Image Processing 5(1): 56–70.

Coggins, J. M. 1983. A framework for texture analysis based on spatial filtering. Ph.D. thesis.

Conrad, S. (ed.). 2000. Petkovic, M. and Riedel, H. (eds.). Content-based Video Retrieval, 74–77.

Costa, L. F. and Cesar, R.M., Jr. 2001. Shape Analysis and Classification: Theory and Practice. CRC Press, Boca Raton, FL, USA.

Danielsson, P. 1980. Euclidean distance mapping. Computer Graphics and Image Processing 14: 227–248.

Datta, R., Joshi, D., Li, J. and Wang, J. Z. 2008. Image retrieval: Ideas, influences, and trends of the new age. ACM Comput. Surv. 40, 2, Article 5.

Davies, E. 1997. Machine Vision: Theory, Algorithms, Practicalities. Academic Press, New York.

Davies, E. R. 2005. Machine Vision: Theory, Algorithms, Practicalities, Morgan Kaufmann, San Francisco.

de Valois, R. L. and De Valois, K. K. 1975. Neural coding of color. pp. 117–166. In: Carterette, E.C., Friedman, M.P. (eds.). Handbook of Perception. Volume 5, Chapter 5, Academic Press, New York, N.Y..

de Valois, R. L. and De Valois, K. K. 1993. A multistage color model. Vision Research 33(8): 1053–1065.

di Baja, G.S. and Thiel, E. 1996. Skeletonization algorithm running on path-based distance maps. Image Vision Computer 14: 47–57.

Dubinskiy, A. and Zhu, S. C. 2003. A Multi-scale Generative Model for Animate Shapes and Parts. In: Proc. Ninth IEEE International Conference on Computer Vision, ICCV.

Dubuisson, M.P. and Jain, A.K. 1994. A modified Hausdorff distance for object matching. Proceedings of the 12th IAPR International Conference on Pattern Recognition, Conference A: Computer Vision & Image Processing (ICPR '94) 1: 566–568.

Duda, R. O., Hart, P. E. and Stork, D. G. 2001. Pattern Classification, second edition, John Wiley & Sons, NY.

Fairchild, M. D. 1998. Color Appearance Models. Addison-Wesley, Readings, MA.

Faloutsos, C. et al. 1994. Efficient and effective querying by image content," Journal of intelligent information systems 3: 231–262.

Fasano, G. and Franceschini, A. 1987. A multidimensional version of the Kolmogorov Smirnov test. Mon. Not. Roy. Astron. Soc. 225: 155–170.

Finlayson, G. D. 1996. Color in perspective. IEEE Trans on Pattern Analysis and Machine Intelligence 8(10): l034–1038, Oct. 1996.

Flusser, J., Suk, T. and Zitová, B. 2009. Moments and Moment Invariants in Pattern Recognition, John Wiley & Sons, Ltd.

Flusser, J.1992. Invariant Shape Description and Measure of Object Similarity. In: Proc. 4th International Conference on Image Processing and its Applications, pp. 139–142.

Foley, J.D., van Dam, A., Feiner, S.K. and Hughes, J.F. 1990. Fundamentals of Interactive Computer Graphics. Addison Wesley, Reading, MA.

Francos, J. M., Meiri, A. A. and Porat, B. 1993. A unified texture model based on a 2d Wold like decomposition," IEEE Trans on Signal Processing pp. 2665–2678, Aug. 1993.

Francos, J. M., Narasimhan, A. and Woods. J. W. 1995. Maximum likelihood parameter estimation of textures using a Wold-decomposition based model. IEEE Trans. on Image Processing, pp. 1655–1666, Dec. 1995.

Franti, P. 2001. Digital Image Processing, Univ. of Joensuu, Dept. of Computer Science, Lecture notes.

Furht, B., Smoliar, S. W. and Zhang, H. J. 1995. Video and Image Processing in Multimedia Systems, Kluwer Academic Publishers.

Gary, J. E. and Mehrotra, R. 1992. Shape similarity-based retrieval in image database systems. Proc. of SPIE, Image Storage and Retrieval Systems 1662: 2–8.

Gevers, T. and Smeulders, A. W. M. 1999. Content-based image retrieval by viewpoint-invariant image indexing. Image and Vision Computing 17(7): 475–488.

Gong, Y., Zhang, H. 1. and Chua, T. C. 1994. An image database system with content capturing and fast image indexing abilities. Proc. IEEE International Conference on Multimedia Computing and Systems, Boston, pp. l21–130, 14–19 May 1994.

Gonzalez, R. C. and Woods, R. E. 2018. Digital Image Processing, fourth edition, Pearson, NY.

Goutsias, J, Vincent, L. and Bloomberg, D. S. (eds.). 2000. Mathematical Morphology and Its Applications to Image and Signal Processing, Kluwer Academic Publishers, Boston, MA.

Grosky, W. I. and Mehrotra, R. 1990. Index based object recognition in pictorial data management," CVGIP 52(3): 416–436.

Gudivada, V. N. and Raghavan, V. V. 1995. Design and evaluation of algorithms for image retrieval by spatial similarity. ACM Trans. on Information Systems 13(2): 115–144, April 1995.

Guo, F., lin, J. and Feng, D. 1998. Measuring image similarity using the geometrical distribution of image contents. Proc. of ICSP, pp. 1108–1112.

Gupta, A. and Jain, R. 1997. Visual information retrieval. Communication of the ACM 40(5): 71–79, May, 1997.

Guru, D. and Nagendraswam, H. 2007. Symbolic representation of two-dimensional shapes. Pattern Recognition Letters 28: 144–155.

Hafner, J. et al. 1995. Efficient color histogram indexing for quadratic form distance functions. IEEE Trans. on Pattern Analysis and Machine Intelligence 17(7): 729–736, July 1995.

Hafner, J., Sawhney, H. S., Equitz, W., Flickner, M. and Niblack, W. 1995. Efficient color histogram indexing for quadratic form distance functions. IEEE Transactions on Pattern Analysis and Machine Intelligence 17(7): 729–735.

Hague, G. E., Weeks, A. R. and Myler, H.R. 1995. Histogram equalization of 24 bit color images in the color difference color space. Journal of Electronic Imaging 4(1): 15–23.

Hall, R. A. 1981. Illumination and Color in Computer Generated Imagery. Springer Verlag, New York, N.Y.

Han, S.and Yang, S. 2005. An Invariant Feature Representation for shape Retrieval. In: Proc. Sixth International Conference on Parallel and Distributed Computing, Applications and Technologies.

Haralick, R. M. 1979. Statistical and structural approaches to texture. Proceedings of the IEEE 67: 786–804.

Haralick, R. M. and Shapiro, L. G. 1992. Computer and Robot vision, Vols. 1 and 2, Addison Wesley, Reading, MA.

Hawkins, J. K. 1970. Textural properties for pattern recognition. Picture Processing and Psychopictorics, pp. 347–370.

Higham Desmond J. and Higham Nicholas, J. 2016. MATLAB Guide, Third Edition, SIAM publications.

Hill, B., Roer, T. and Vorhayen, F.W. 1997. Comparative analysis of the quantization of color spaces on the basis of the CIE-Lab color difference formula. ACM Transaction of Graphics 16(1): 110–154.

Hitchcock, F. L. 1941. The distribution of a product from several sources to numerous localities. Journal of Math. Phys. 20: 224–230.

Holla, K. 1982. Opponent colors as a 2-dimensional feature within a model of the first stages of the human visual system. Proceedings of the 6th Int. Conf. on Pattern Recognition 1: 161–163.

Hollink, L., Schreiber, G., Wielinga, B. and Worring, M. 2004.Classification of user image descriptions", International Journal of Human Computer Studies 61: 601–626.

http://homepages.inf.ed.ac.uk/rbf/BOOKS/PHILLIPS/.

http://mpeg7.org/visual-descriptors/.

https://en.wikipedia.org/wiki/Digital_image_processing.

https://www.mathworks.com/discovery/digital-image-processing.html.

Hu, K. 1977. Visual pattern recognition by moment invariants. In: Aggarwal, J. K., Duda, R. O. and Rosenfeld, A. (eds.). Computer Methods in Image Analysis. IEEE computer Society, Los Angeles, CA.

Hu, M. K.1962. Visual Pattern Recognition by Moment Invariants. IRE Trans. Information Theory IT-8: 179–187.

Huang, J. et al. 1997. Image indexing using color correlogram. IEEE Int. Conf. on Computer Vision and Pattern Recognition, pp. 762–768, Puerto Rico.

Huang, J., Kumar, S. R. and Metra, M. 1997. Combining supervised learning with color correlograms for content-based image retrieval. Proc. of ACM Multimedia 95: 325–334.

Huang, J., Kumar, S. R., Mitra, M., Zhu, W.-J. and Zabih, R. 1997. Image indexing using color correlograms. Proceedings of Computer Vision and Pattern Recognition, pp. 762–768.

Huang, J., Kumar, S.R. Metra, M., Zhu, W. J. and Zabith, R. 1999. Spatial color indexing and applications," Int'l J Computer Vision 35(3): 245–268.

Hurvich, Leo M. 1981. Color Vision. Sinauer Associates, Sunderland MA.

Huttenlocher, D.P., Klanderman, G.A. and Rucklidge, W.J. 1993. Comparing images using the Hausdorff distance. IEEE Transactions on Pattern Analysis and Machine Intelligence 15(9): 850–863.

Iivarinen, J. and Visa, A. 1996. Shape recognition of irregular objects. In: Proc. SPIE, Intelligent Robots and Computer Vision XV: Algorithms, Techniques, Active Vision, and Materials Handling, pp. 25–32.

ISO/IEC JTC1/SC29/WG11, MPEG-7 Overview (version 10), Technical report (2004).

Jagadish, H. V. 1991. A retrieval technique for similar shapes. Proc. of Int. Conf on Management of Data, SIGMOID'9i, Denver, CO, pp. 208–217.

Jahne, B. 2004. Practical Handbook on Image Processing for Scientific and Technical Applications, Second Edition, CRC Press.

Jain, A. K. 1989. Fundamentals of Digital Image Processing, Prentice-Hall.

Jain, A. K. and F. Farroknia. 1991. Unsupervised texture segmentation using Gabor filters. Pattern Recognition 24(l2): 1167–1186.

Jain, P. and Tyagi, V. 2016. A survey of edge-preserving image denoising methods. Inf Syst Front 18(1): 159–170.

Jain, R. 1992. Proc. US NSF Workshop Visual Information Management Systems.

Jain, R., Pentland, A. and Petkovic, D. 1995. Workshop Report: NSF-ARPA Workshop on Visual information Management Systems, Cambridge, Mass, USA.

Jalba, A., Wilkinson, M. and Roerdink, J. 2006. Shape representation and recognition through morphological curvature scale spaces. IEEE Trans. Image Processing 15(2): 331–341.

Jameson, D. and Hurvich, L.M. 1968. Opponent-response functions related to measured cone photo pigments. Journal of the Optical Society of America 58: 429–430.

Jayaraman, S., Esakirajan, S. and Veerakumar, T. 2015. Digital Image Processing, McGraw Hill.

Jin, K., Cao, M., Kong, S. and Lu, Y. 2006. Homocentric Polar-Radius Moment for Shape Classification. In: The 8th International Conference on Proc. Signal Processing.

Johnson, G. M. and Fairchild, M. D. 1999. Full spectral color calculations in realistic color image synthesis. IEEE Computer Graphics and Applications 19(4): 47–53.

Judd, D. B. and Wyszecki, G. 1975. Color in Business, Science and Industry. John Wiley, New York, N.Y.

Julesz, B. 1965. Texture and visual perception. Scientific American 212: 38–49.

Jurman, G., Riccadonna, S., Visintainer, R. and Furlanello, C. 2009. Canberra Distance on Ranked Lists. In Proceedings, Advances in Ranking – NIPS 09 Workshop, pp. 22–27.

Kailath, T. 1967. The divergence and Bhattacharyya distance measures in signal selection. IEEE Trans. Comm. Techn., pp. 52–60.

Kan, C. and Srinath, M. D. 2002. Invariant character recognition with Zernike and orthogonal Fourier-Mellin moments. Pattern Recognition 35: 143–154.

Kasson, M.J. and Ploaffe, W. 1992. An analysis of selected computer interchange color spaces. ACM Transaction of Graphics 11(4): 373–405.

Kauppinen, H., Seppanen, T. and Pietikainen, M. 1995. An Experimental Comparison of Auto-regressive and Fourier-Based Descriptors in 2-D Shape Classification. IEEE Trans. Pattern Analysis and Machine Intelligence 17(2): 201–207.

Kauppinen, H., Seppnaen, T. and Pietikainen, M. 1995. An experimental comparison of autoregressive and Fourier-based descriptors in 20 shape classification. IEEE Trans. Pattern Anal. and Machine intell. 17(2): 201–207.

Khalil, M. and Bayoumi, M. 2001. A Dyadic Wavelet Affine Invariant Function for 2D Shape Recognition. IEEE Trans. Pattern Analysis and Machine Intelligence 25(10): 1152–1164.

Kolesnikov, A. 2003. Efficient algorithms for vectorization and polygonal approximation, Ph.D thesis, University of Joensu, Finland.

Kpalma, K. and Ronsin, J. 2006. Multiscale contour description for pattern recognition. Pattern Recognition Letters 27: 1545–1559.

Krzanowski. 1995. Recent Advances in Descriptive Multivariate Analysis, Chapter 2, Oxford science publications.

Kullback, S. and Leibler, R. A. 1951. On information and sufficiency. Annals of Mathematical Statistics 22(1): 79–86.

Kullback, S. 1968. Information Theory and Statistics. Dover Publications, New York.

Kumar, A., Ghrera, S.P. and Tyagi, V. 2015. A Comparison of Buyer-Seller Watermarking Protocol (BSWP) Based on Discrete Cosine Transform (DCT) and Discrete Wavelet Transform (DWT), Advances in Intelligent Systems and Computing, vol 337. Springer.

Lance, G. N. and Williams, W. T. 1966. Computer programs for hierarchical polythetic classification ("similarity analysis"). Computer Journal, Vol. 9(1): 60–64.

Latecki, L. J. and Lakamper, R. 1999. Convexity rule for shape decomposition based on discrete Contour Evolution. Computer Vision and Image Understanding 73(3): 441–454.

Latecki, L. J. and Lakamper, R. 2000. Shape Similarity Measure Based on Correspondence of Visual Parts. IEEE Trans. Pattern Analysis and Machine Intelligence 22(10): 1185– 1190.

Lee, S.-M., Abbott, A. L., Clark, N. A. and Araman, P. A. 2006. A Shape Representation for Planar Curves by Shape Signature Harmonic Embedding. In: Proc. IEEE Computer Society Conference on Computer Vision and Pattern Recognition.

Lee, S. Y. and Hsu, F. H. 1990. 2D C-string: a new spatial knowledge representation for image database systems. Pattern Recognition 23: 1077–1087.

Lee, S. Y., Yang, M.C. and Chen, J. W. 1992. 2D B-string: a spatial knowledge representation for image database system. Proc. ICSC'92 Second into computer Sci. Conf., pp. 609–615.

Levine, M. D. 1985. Vision in man and machine, McGraw-Hill College.

Levkowitz, H. and Herman, G. T. 1993. GLHS: a generalized lightness, hue and saturation color model. Graphical Models and Image Processing, CVGIP 55(4): 271–285.

Li, J., Wang, J. Z. and Wiederhold, G. 2000. IRM: integrated region matching for image retrieval. In Proceedings of the eighth ACM international conference on multimedia, pp. 147–156.

Lin, J. 1991. Divergence measures based on the Shannon entropy. TIT.

Liu, F. and Picard, R. W.1996. Periodicity, directionality, and randomness: Wold features for image modeling and retrieval. IEEE Trans. on Pattern Analysis and Machine Learning 18(7).

Liu, H., Song, D., Rüger, S., Hu, R. and Uren, V. 2008. Comparing Dissimilarity Measures for Content-Based Image Retrieval, Information Retrieval Technology. AIRS 2008. Lecture Notes in Computer Science, vol. 4993. Springer, pp. 44–50.

Liu, Y. et al. 2007. A survey of content-based image retrieval with high-level semantics, Pattern Recognition 40: 262–282.

Liu, Y., Zhang, D. and Lu, G. 2007. A survey of content-based image retrieval with high-level semantics. Pattern Recognition 40: 262–282.

Liu, Y. K., Wei, W., Wang, P. J. and Zalik, B. 2007. Compressed vertex chain codes. Pattern Recognition 40(11): 2908–2913.

Long, F., Zhang, H. and Feng D. D. 2003. Fundamentals of Content-Based Image Retrieval. In: Feng, D. D., Siu, W. C. and Zhang, H. J. (eds.). Multimedia Information Retrieval and Management. Signals and Communication Technology. Springer, Berlin, Heidelberg

Lotufo, R. A., Falcao, A. A. and Zampirolli, F. A. 2000. Fast Euclidean distance transform using a graph-search algorithm. In Proceeding of SIBGRAPI, pp. 269–275.

Lu, Guoyun. 1996. Communication and Computing for Distributed Multimedia Systems. Artech House Publishers, Boston, MA.

Lu, G. and Sajjanhar, A. 1999. Region-based shape representation and similarity measure suit-able for content based image retrieval. ACM Multimedia System Journal 7(2): 165–174.

Lu, K.-J. and Kota, S. 2002. Compliant Mechanism Synthesis for Shape-Change Applications: Preliminary Results. pp. 161–172. In: Proceedings of SPIE Modelling, Signal Processing, and Control Conference.

Luo, J., Joshi, D., Yu, J. and Gallagher, A. 2011. Geotagging in multimedia and computer vision-a survey, Multimedia Tools and Applications 51(1): 187–211.

Luong, Q. T. 1993. Color in computer vision. Handbook of Pattern Recognition and Computer Vision, Word Scientific Publishing Company 311–368.

Ma, W. Y. and Manjunath, B. S. 1995. A comparison of wavelet features for texture annotation. Proc. of IEEE Int. Conf. on Image Processing, Vol. II, pp. 256–259, Washington D.C.

Ma, W. Y. and Manjunath, B. S. 1995. Image indexing using a texture dictionary. Proc. of SPIE Conf. on image Storage and Archiving System, Vol. 2606, pp. 288–298, Philadelphia, Pennsylvania.

Ma, W. Y. and Manjunath, B. S. 1997. Edge flow: a framework of boundary detection and image segmentation. IEEE into Conf. on Computer Vision and Pattern Recognition, pp. 744–749, Puerto Rico.

MacDonald, L. W. 1999. Using color effectively in computer graphics. IEEE Computer Graphics and Applications 19(4): 20–35.

Mahalanobis, P. C.1936. On the generalised distance in statistics. Proceedings of the National Institute of Sciences of India 2(1): 49–55.

Mallat, S. G. 1989. A theory for multiresolution signal decomposition: the wavelet representation. IEEE Trans. Pattern Analysis and Machine Intelligence, Vol. 11, pp. 674–693.

Manjunath, B. S. and Ma, W. Y. 1996. Texture features for browsing and retrieval of image data. IEEE Trans. on Pattern Analysis and Machine intelligence 18(8): 837–842.

Manning, C. D. 2009. Prabhakar Raghavan & Hinrich Schütze, An introduction to information retrieval, Cambridge University press.

Mao, J. and Jain, A. K. 1992. Texture classification and segmentation using multiresolution simultaneous autoregressive models. Pattern Recognition 25(2): 173–188.

Marques, Oge. 2011. Practical Image and Video Processing Using MATLAB, Wiley IEEE press.

Mathias, E. 1998. Comparing the influence of color spaces and metrics in content-based image retrieval. Proceedings of International Symposium on Computer Graphics, Image Processing, and Vision, pp. 371–378.

Maxwell, J. C. 1890. On the theory of three primary colors. Science Papers Cambridge University Press: 445–450.

McLaren, K. 1976. The development of the CIE L*a*b* uniform color space. J. Soc. Dyers Color, 338–341.

Mehtre, B. M., Kankanhalli, M. S. and Lee, W. F. 1997. Shape Measures for Content Based Image Retrieval: A Comparison. Pattern Recognition 33(3): 319–337.

Meijster, A., Roerdink, J. and Hesselink, W. 2000. A general algorithm for computing distance transforms in linear time. In Math. Morph. and Its Appls. to Image and Signal Proc., pp. 331–340.

Miano, J. 1999. Compressed Image File Formats: JPEG, PNG, GIF, XBM, BMP, ACM Press/Addison-Wesley Publ. Co., New York, NY, USA.

Miguel, Arevalillo-Herráez, Juan Domingo, Francesc J. Ferri. 2008. Combining similarity measures in content-based image retrieval, Pattern Recognition Letters 29(16): 2174–2181.

Mingqiang Yang, Kidiyo Kpalma, Joseph Ronsin. 2012. Shape-based invariant features extraction for object recognition. Advances in reasoning-based image processing, analysis and intelligent systems: Conventional and intelligent paradigms, Springer.

Mokhtarian, F. and Mackworth, A. K. 1992. A Theory of Multiscale, Curvature-Based Shape Representation for Planar Curves. IEEE Trans. Pattern Analysis and Machine Intelligence 14(8): 789–805.

Mori, G. and Malik, J. 2002. Estimating Human Body Configurations Using Shape Context Matching. pp. 666–680. In: Heyden, A., Sparr, G., Nielsen, M. and Johansen, P. (eds.). ECCV 2002. LNCS, vol. 2352. Springer, Heidelberg.

MPEG Video Group, Description of core experiments for MPEG-7 color/texture descriptors, ISO/MPEGJTC1/SC29/WGll/MPEG98/M2819, July 1999.

Mukundan, R. 2004. A new class of rotational invariants using discrete orthogonal moments. pp. 80–84. In: Sixth IASTED International Conference on Signal and Image Processing.

Muller, H. and Muller, W. et al. 2001. Performance evaluation in Content based image retrieval: overview and proposals, Pattern Recognition Letters 22: 593–601.

Myler, H. R. and Arthur R. Weeks. 2009. The pocket handbook of image processing algorithms in C, Prentice Hall Press, NJ.

Niblack, W., Barber, R., Equitz, W., Flickner, M.D., Glasman, E.H., Petkovic, D., Yanker, P., Faloutsos, C., Taubin, G. and Heights, Y. 1993. Querying images by content, using color, texture, and shape. In SPIE Conference on Storage and Retrieval for Image and Video Databases 1908: 173–187.

Nixon, M.S. and Aguado, A.S. 2002. Feature Extraction and Image Processing, Newnes Publishers.

Ohta, Y., Kanade, T. and Sakai, T. 1980. Color information for region segmentation. Computer Graphics and Image Processing 13: 222–241.

Ojala, T., Pietikainen, M. and Harwood, D. 1996. A comparative study of texture measures with classification based feature distributions. Pattern Recognition 29(1): 51–59.

Osterreicher, F. and Vajda, I. 2003. A new class of metric divergences on probability spaces and its statistical applications. AISM.

Paglieroni, D. W. 1992. Distance transforms: properties and machine vision applications. CVGIP: Graphical Models and Image Processing 54(1): 56–74.

Palus, H. 1998. Color spaces. pp. 67–89. In: Sangwine, S.J. and Horne, R.E.N. (eds.) The Color Image Processing Handbook, Chapman & Hall, Cambridge, Great Britain.

Pass, G. and Zabith, R. 1996. Histogram refinement for content-based image retrieval. IEEE Workshop on Applications of Computer Vision, pp. 96–102.

Pass, G. and Zabith, R. 1999. Comparing images using joint histograms. Multimedia Systems 7: 234–240.

Pele, O. and Werman, M. 2010. The Quadratic-Chi Histogram Distance Family, in Lecture Notes of Computer Science 6312: 749–762.

Peng, J., Yang, W. and Cao, Z. 2006. A Symbolic Representation for Shape Retrieval in Curvature Scale Space. In: Proc. International Conference on Computational Intelligence for Modelling Control and Automation and International Conference on Intelligent Agents Web Technologies and International Commerce.

Persoon, E. and Fu, K. 1977. Shape discrimination using Fourier descriptors. IEEE Trans. Syst. Man and Cybern. 7: 170–179.

Petrou, M. and Petrou, C. 2010. Image Processing: The Fundamentals, John Wiley & Sons, NY.

Picard, R. W., Kabir, T. and Liu, F. 1993. Real-time recognition with the entire Brodatz texture database. Proc. IEEE Int. Conf. on Computer Vision and Pattern Recognition, pp. 638–639, New York.

Plataniotis, K. N. and Venetsanopoulos, A. N. 2000. Color Spaces. In: Color Image Processing and Applications, Digital Signal Processing, Springer, Berlin, Heidelbergm.

Poynton, C. A. A Technical Introduction to Digital Video. Prentice Hall, Toronto.

Pratt, W. K. 2014. Introduction to Digital Image Processing, CRC Press, Boca Raton, FL.

Rangayyan, R. M. 2005. Biomedical Image Analysis, CRC Press, Boca Raton, FL.

Rhodes, P.A. 1998. Color management for the textile industry. Sangwine, S.J., Horne, R.E.N. (eds.). The Color Image Processing Handbook, 307–328, Chapman & Hall, Cambridge, Great Britain.

Ricard, J., Coeurjolly, D. and Baskurt, A. 2005. Generalizations of angular radial transform for 2D and 3D shape retrieval. Pattern Recognition Letters 26(14).

Richards, W. and Polit, A. 1974. Texture matching. Kybernetik 16:155–162.

Robertson, P. and Schonhut, J. 1999. Color in computer graphics. IEEE Computer Graphics and Applications 19(4): 18–19.

Robinson, J. T. 1981. The k-d-B-tree: a search structure for large multidimensional dynamic indexes. Proc. of SIGMOD Conference, Ann Arbor.

Rosenfield, A. and Kak, A. C. 1982. Digital picture processing, second edition, vols. 1 & 2, Academic press, New York.

Rubner, Y. 1999. Perceptual Metrics for Image Database Navigation. PhD Thesis, Stanford University, May 1999.

Rui, Hu, Stefan, Ruger, Dawei, Song, Haiming, Liu and Zi Huang. 2008. Dissimilarity Measures for Content-Based Image Retrieval. IEEE International Conference on Multimedia and Expo, pp. 1365–1368.

Rui, Y., Huang, T. S., Ortega, M. and Mehrotra, S. 1998. Relevance feedback: a power tool for interactive content-based image retrieval. IEEE Trans. on Circuits and Systems for Video Technology.

Rui, Y., Huang, T.S. and Mehrotra, S. 1997. Content-based image retrieval with relevance feedback in MARS. Proceedings of International Conference on Image Processing 2: 815–818.

Russ, J. C. 2011. The Image Processing Handbook, 6th edition, CRC Press, Boca Raton, FL.

Safar, M., Shahabi, C. and Sun, X. 2000. Image Retrieval by Shape: A Comparative Study. In IEEE International Conference on Multimedia and Expo (I), pp. 141–144.

Saito, T. and Toriwaki, J. I. 1994. New algorithms for Euclidean distance transformation of an n dimensional digitized picture with application. Pattern Recognition 27(11): 1551–1565.

Salton, G. and McGill, M. 1983. Introduction to Modern Information Retrieval. McGraw-Hill, New York, NY.

Sclaroff, S., Taycher, L. and Cascia, M. L. 1997. ImageRover: a content-based image browser for the World Wide Web. Boston University CS Dept. Technical Report 97-005.

Sebastian, T., Klein, P. and Kimia, B. 2004. Recognition of Shapes by Editing Their Shock Graphs. IEEE Trans. Pattern Analysis and Machine Intelligence 26(5): 550–571.

Shapiro, L. G. and Stockman, G. C. 2001. Computer Vision, Prentice-Hall, Upper Saddle River, NJ.

Sharma, G. and Trussel, H. J. 1997. Digital color processing. IEEE Trans. on Image Processing, 6(7): 901–932.

Sharma, G., Yrzel, M. J. and Trussel, H. J. 1998. Color imaging for multimedia. Proceedings of the IEEE 86(6): 1088–1108.

Shashank, J., Kowshik, P., Srinathan, K. and Jawahar, C. V. 2008. Private Content Based Image Retrieval. 2008 IEEE Conference on Computer Vision and Pattern Recognition, Anchorage, AK, pp. 1–8.

Shih, Tian-Yuan. 1995. The reversibility of six geometric color spaces. Photogrammetric Engineering and Remote Sensing 61(10): 1223–1232.

Siddiqi, K. and Kimia, B. 1996. A Shock Grammar for Recognition. In: Proceedings of the IEEE Conference Computer Vision and Pattern Recognition, pp. 507–513.

Sklansky, J. 1978. Image Segmentation and Feature Extraction," {IEEE} Trans. Systems, Man, and Cybernetics 8: 237–247.

Smith, A. R. 1978. Color gamut transform pairs. Computer Graphics (SIGGRAPH'78 Proceedings) 12(3): 12–19.

Smith, S. P. and Jain, A. K. 1982. Chord distribution for shape matching. Computer Graphics and Image Processing 20: 259–271.

Soile, P. 2003. Morphological image analysis: Principles and applications, second edition, Springer-Verlag, New York.

Soille, P. 2003. Morphological Image Analysis: Principles and Applications, second edition, Springer-Verlag, NY.

Sonka, M., Hlavac, V. and Boyle, R. 2008. Image Processing, Analysis and Machine Vision, Third edition, Thomson Learning.

Stricker, M. and Orengo, M. 1995. Similarity of color images. In SPIE Conference on Storage and Retrieval for Image and Video Databases III 2420: 381–392.

Sundararajan, D. 2017. Digital Image Processing: A Signal Processing and Algorithmic Approach, Springer Singapore.

Swain, M. J. and Ballard, D. H. 1991. Color indexing. International Journal of Computer Vision 7(1): 11–32.

Tamura, H., Mori, S. and Yamawaki, T. 1978. Textural features corresponding to visual perception. IEEE Transactions on Systems, Man, and Cybernetics 8: 460–473.

Taubin, G. and Cooper, D. 1991. Recognition and positioning of rigid objects using algebraic moment invariants. In: SPIE Conference on Geometric Methods in Computer Vision, pp. 175–186.

Taza, A. and Suen, C. 1989. Discrimination of planar shapes using shape matrices. IEEE Trans. System, Man, and Cybernetics 19(5): 1281–1289.

Tegolo, D. 1994. Shape analysis for image retrieval. Proc. of SPIE. Storage and Retrieval for Image and Video Databases-II, no. 2185, San Jose, CA, pp. 59–69, February 1994.

Tektronix, TekColor Color Management System: System Implementers Manual. Tektronix Inc., 1990.

Thayananthan, A., Stenger, B., Torr, P. H. S. and Cipolla, R. 2003. Shape Context and Chamfer Matching in Cluttered Scenes. In: Proc. IEEE Computer Society Conference on Computer Vision and Pattern Recognition.

Tieng, Q. M. and Boles, W. W. 1997. Wavelet-Based Affine Invariant Representation: A Tool for Recognizing Planar Objects in 3D Space. IEEE Trans. Pattern Analysis and Machine Intelligence 19(8): 846–857.

Tiwari, D. and Tyagi, V. 2017. Digital Image Compression, Indian Science Cruiser, Vol. 31 (6): 44–50.

Tominaga, S. 1986. Color image segmentation using three perceptual attributes. Proceedings of CVPR'86, 1: 628–630.

Torres, R. S. and Falcão, A. X. 2006. Content-Based Image Retrieval: Theory and Applications. RITA, Vol. XIII (2), DOI: 10.1.1.89.34.

Trussell, H. J and Vrhel, M. J. 2008. Fundamentals of Digital Imaging, Cambridge University Press, UK.

Tsai, C.-F. and Hung, C. 2008. Automatically annotating images with keywords: A Review of Image Annotation Systems. Recent Patents on Computer Science 1: 55–68.

Tyagi, Vipin. 2017. Content-Based Image Retrieval: Ideas, Influences, and Current Trends, Springer Singapore Publisher.

Umbaugh, S. E. 2005. Computer Imaging: Digital Image Analysis and Processing, CRC Press, Boca Raton, FL.

Umbaugh, S. E. 2010. Digital Image Processing and Analysis, second edition, CRC Press, Boca Raton, FL.

Veltkamp, R. C. and Hagedoorn, M. 1999. State-of-the-art in shape matching. Technical Report UU-CS-1999-27, Utrecht University, Department of Computer Science, Sept.

Vendrig, J., Worring, M. and Smeulders, A. W. M. 1999. Filter image browsing: exploiting interaction in retrieval. Proc. Viusl'99: Information and Information System.

von Stein, H.D. and Reimers, W. 1983. Segmentation of color pictures with the aid of color information and spatial neighborhoods. Signal Processing II: Theories and Applications 1: 271–273.

Voorhees, H. and Poggio, T. 1988. Computing texture boundaries from images. Nature 333: 364–367.

Wang, B., Zhang, X., Wang, M. and Zhao, P. 2010. Saliency distinguishing and applications to semantics extraction and retrieval of natural image. International Conference on Machine Learning and Cybernetics (ICMLC) 2: 802–807.

Wang, H., Guo, F., Feng, D. and Jin, J.1998. A signature for content-based image retrieval using a geometrical transform. Proc. of ACM MM'98, Bristol, UK.

Wang, J. Z. 2001. Integrated Region-Based Image Retrieval, Kluwer Academic Publishers.

Wang, Y. P. and Lee, K. T. 1999. Multiscale curvature-based shape representation using B-spline wavelets. IEEE Trans. Image Process 8(10): 1586–1592.

Weeks, A. R. 1996. Fundamentals of Electronic Image Processing. SPIE Press, Piscataway, New Jersey.

Weng, L., Amsaleg, L., Morton, A. and Marchand-Maillet, S. 2015. A Privacy-Preserving Framework for Large-Scale Content-Based Information Retrieval. In IEEE Transactions on Information Forensics and Security 10(1): 152–167. doi: 10.1109/TIFS.2014.2365998.

Wong, W. H., Siu, W. C. and Lam, K. M. 1995. Generation of moment invariants and their uses for character recognition. Pattern Recognition Letters 16: 115–123.

www.mathworks.com.

Wyszecki, G. and Stiles, W. S. 1996. Color Science, Concepts and Methods, Quantitative Data and Formulas. John Wiley, N.Y., 2nd Edition.

Yang, L. and Aigregtsen, F. 1994. Fast computation of invariant geometric moments: A new method giving correct results. Proc. IEEE Int. Conf. on Image Processing.

Yang, M., Kpalma, K. and Ronsin, J. 2007. Scale-controlled area difference shape descriptor. In: Proc. SPIE, Electronic Imaging science and Technology.

Young, T. 1802. On the theory of light and colors. Philosophical Transactions of the Royal Society of London 92: 20–71.

Zahn, C. T. and Roskies, R. Z. 1972. Fourier Descriptors for Plane closed Curves. IEEE Trans. Computer c-21(3): 269–281.

Zhang, D. and Lu, G. 2002. A Comparative Study of Fourier Descriptors for Shape Representation and Retrieval. In: Proc. 5th Asian Conference on Computer Vision.

Zhang, D. and Lu, G. 2003. A comparative study of curvature scale space and Fourier descriptors for shape-based image retrieval. Visual Communication and Image Representation 14(1).

Zhang, D. and Lu, G. 2004. Review of shape representation and description techniques. Pattern Recognition 37: 1–19

Zhang, D. S. and Lu, G. 2001. A Comparative Study on Shape Retrieval Using Fourier Descriptors with Different Shape Signatures. In: Proc. International Conference on Intelligent Multimedia and Distance Education (ICIMADE 2001).

Zhang, H. and Malik, J. 2003. Learning a discriminative classifier using shape context distances. In: Proc. IEEE Computer Society Conference on Computer Vision and Pattern Recognition.

Zhang, H. J. and Zhong, D. 1995. A scheme for visual feature-based image indexing. Proc. of SPIE conf. on Storage and Retrieval for Image and Video Databases III, pp. 36–46, San Jose.

Zhang, H. J. et al. 1995. Image retrieval based on color features: An evaluation study. SPIE Conf. on Digital Storage and Archival, Pennsylvania.

Zhou, X. S. and Huang, T. S. 2000. CBIR: from low-level features to high- level semantics, Proceedings of the SPIE, Image and Video Communication and Processing, San Jose, CA 3974: 426–431.

Index

Color Figures Section

Chapter 1

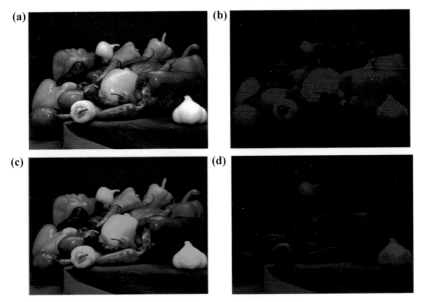

Fig. 1.1. (a), (b), (c), (d)

Fig. 1.4 (a)

Chapter 2

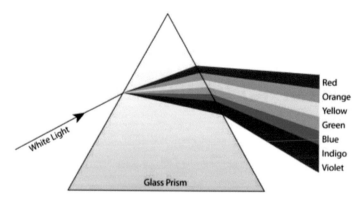

Fig. 2.5

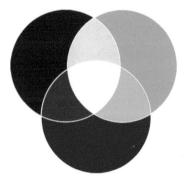

Fig. 2.6

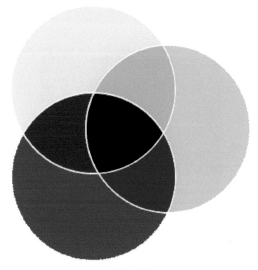

Fig. 2.8

Chapter 4

Fig. 4.5 (a)

Fig. 4.10 (a)

Chapter 7

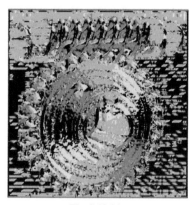

Fig. 7.13 (c)

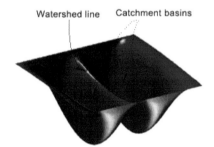

Fig. 7.16

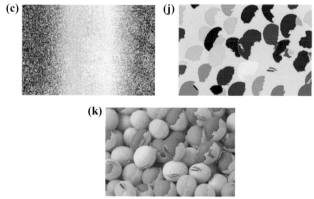

Fig. 7.17 (c), (j), (k)